STANDARD
OIL ABBREVIATOR

Second Edition

Compiled by
the Association of
Desk and Derrick Clubs

PennWell Books
Division of
PennWell Publishing Company
Tulsa, Oklahoma

International Standard Book Number:
0-87814-017-4

Library of Congress
Catalog Card Number:
72-96172

Printed in the United States of America

4 5   82 81

Preface to the Second Edition

The gratifying response to the first edition of the D&D Oil Abbreviator, which carried it through three printings, seemed to demand a second edition.

As in all undertakings, there are changes that "would be made if we had it to do over."

So the ladies of Desk and Derrick got behind the project a second time and submitted more data, pointed out earlier typographical errors, and suggested new features for this second edition.

This volume contains more abbreviations for words commonly used in the oil and gas industries than its predecessor. It also contains more abbreviations for company, association, agency, and society names. These are separated geographically, in a sense, with one list covering the U.S. and Canada and the second encompassing the rest of the world.

A major change is noted in the handling of logging tools and services. Because standards have not been established, and because abbreviations for these tools and services are rather complex, it was decided to contain them under a separate heading.

And as a convenience for secretaries, engineers, draftsmen, and others in the industry, the back of the book contains standard map symbols through the courtesy of API, mathematical symbols, and the Greek alphabet.

ALL in all, this second edition was designed to carry out the hope of the first—to fill a need.

Preface to the First Edition

This book was developed from an idea proposed by Bettye Lynn Hatcher, Monahans, Texas, when she was president of the Monahans Desk and Derrick Club. She presented the idea in a feature story in the club's publication, "The Direct Line."

Her story described the plight of a young secretary on her first assignment with an oil company, trying to decipher the strange-looking language in a drilling report. With no "key" to unlock the mystery of this language, she was completely lost.

So here is the "key"—a listing of nearly every abbreviation used in the oil and gas industry, supplied by those most familiar with this peculiar brand of shorthand—the members of the Association of Desk and Derrick Clubs of North America.

Contributions came from these members in all geographical areas and in all branches of the industry from exploration to marketing. Every effort was made to make this first product as complete as possible, but omissions are bound to occur. Additional contributions are invited by The Oil and Gas Journal's Book Department.

For convenience, the book is divided into sections. The first two of these are the largest and give abbreviations for words or phrases. The first section shows the abbreviation followed by its definition. In the second section this order is reversed.

The third section gives the American abbreviations for oil-related companies operating in countries other than the United States. This is followed by abbreviations for the many associations and societies related to the oil industry.

At the outset, the reader will note that there are some abbreviations that apply to more than one word or phrase. Out of context, this seems confusing, but when used in context, these abbreviations take on a clearer meaning and are easily understood by those familiar with the industry's everyday language.

On the matter of style, the contributions came in assorted forms; all capital letters, all lower-case letters, mixed caps and lower-case, etc. This is not surprising because there are no standards to follow, yet, and each secretary is her own boss in this regard. But to introduce some consistency, the editors adopted a style based on simple logic: If an abbreviation is made up of the principal letters in a word, as in *bldg* for *building*, lower-case letters are used. If an abbreviation is made up of the first letters of several words (*WOC* for *waiting on cement*), capital letters are used. And to make the abbreviations as abbreviated as possible, periods are used only when their absence could lead to confusion.

It is hoped that this book fills a need in the industry, not only to help the secretarial side interpret the language, but to familiarize all newcomers to the industry with our peculiar ways of doing things.

What is D & D

Conceived by an ambitious secretary, nurtured by women with vision, encouraged by cooperative employers and a progressive industry—this is the saga of DESK and DERRICK. It is a unique organization of six thousand women employed in the petroleum and allied industries. All one hundred clubs, spread over the North American continent, are dedicated to the proposition that GREATER KNOWLEDGE of the petroleum industry will result in GREATER SERVICE in job performance.

From New Orleans, where the first club was formed in April 1949, the idea spread in the ensuing eighteen months to Jackson, Mississippi, Los Angeles, and Houston; in September 1962 in Houston the "Queen of Clubs" held its first annual convention.

The Desk and Derrick purpose is implemented by educational monthly programs, by field trips to industry installations and by special study courses.

Nonshareholding, noncommercial, nonprofit, nonpartisan, and nonbargaining in its policies, the organization nevertheless has very positive concepts on the value of education for women.

Contents

A

A/	Acidized with
AA	After acidizing, as above
A&A	Adjustments and Allowances
ab	Above
ABC	Audit Bureau of Circulation
abd	Abandoned
abd loc	Abandoned location
abd-gw	Abandoned gas well
abd-ow	Abandoned oil well
abdogw	Abandoned oil & gas well
abrsi jet	Abrasive jet
ABS	Acrylonitrile butadiene styrene rubber
absrn	Absorption
abst	Abstract
abt	About
abun	Abundant
abv	Above
ac	Acid, acidizing
ac	Acres, acre, acreage
AC	Alternating current
AC	Austin Chalk
acct	Account, account of, accounting
accum	Accumulative
acd	Acidize, acidizing, acidized
A-Cem	Acoustic cement
acfr	Acid fracture treatment
ac-ft	Acre feet
ACM	Acid-cut mud
acrg	Acreage
ACSR	Aluminum conductor steel reinforced
ACT	Automatic custody transfer

1

ACW	Acid-cut water
AD	Authorized depth
add	Additive
addl	Additional
adj	Adjustable
adm	Administrative, administration
ADOM	Adomite
ADP	Automatic data processing
adpt	Adapter
adspn	Adsorption
advan	Advanced
AF	Acid frac, after fracture
AFE	Authorization for expenditure
affd	Affirmed
afft	Affidavit
AFP	Average flowing pressure
aggr	Aggregate
aglm	Agglomerate
AIR	Average injection rate
Alb	Albany
alg	Algae
alk	Alkalinity
alky	Alkylate, alkylation
allow	Allowable
alm	Alarm
alt	Alternate
amb	Ambient
amor	Amorphous
amort	Amortization
amp	Ampere
amph	Amphipore
Amph	Amphistegina
amp hr	Ampere hour
amt	Amount
anal	Analysis, analytical
ang	Angle, angular
Angul	Angulogerina
anhy	Anhydrite, anhydritic
anhyd	Anhydrous
ANYA	Allowable not yet available

AOF	Absolute open flow potential (gas well)
app	Appears, appearance
appd	Approved
appl	Appliance, applied
applic	Application
approx	Approximate(ly)
apr	Apparent(ly)
apt	Apartment
aq	Aqueous
AR	Acid residue
A/R	Accounts receivable
Ara	Arapahoe
arag	Aragonite
Arb	Arbuckle
arch	Architectural
Archeo	Archeozoic
aren	Arenaceous
arg	Argillaceous, argillite
ark	Arkose(ic)
Arka	Arkadelphia
arm	Armature
arnd	Around
ARO	At rate of
arom	Aromatics
AS	Anhydrite stringer, after shot
ASAP	As soon as possible
asb	Asbestos
asbr	Absorber
asgmt	Assignment
Ash	Ashern
asph	Asphalt, asphaltic
assgd	Assigned
assn	Association
assoc	Associate (d) (s)
asst	Assistant
assy	Assembly
astn	Asphaltic stain
AS&W ga	American Steel & Wire gauge
at	Atomic

AT	All thread, acid treat(ment), after treatment
At	Atoka
ATC	After top center
ATF	Automatic transmission fluid
atm	Atmosphere, atmospheric
att	Attempt(ed)
atty	Attorney
at wt	Atomic weight
aud	Auditorium
Aus	Austin
auth	Authorized
auto	Automatic, automotive
autogas	Automotive gasoline
aux	Auxiliary
AV	Annular velocity, Aux Vases sand
av	Aviation
avail	Available
AVC	Automatic volume control
avg	Average
avgas	Aviation gasoline
AW	Acid water
AWG	American Wire Gauge
awtg	Awaiting
az	Azimuth
aztrop	Azeotropic

B

B/	Base, bottom of given formation (i.e., B/Frio)
BA	Barrels of acid
Ball	Balltown sand
bar	Barite(ic)
bar	Barometer or barometric
BAR	Barrels acid residue
Bar	Barlow Lime

Bark Crk	Barker Creek
Bart	Bartlesville
base	Basement (Granite)
bat	Battery
BAT	Before acid treatment
Bate	Bateman
BAW	Barrels acid water
BAWPD	Barrels Acid Water Per Day
BAWPH	Barrels Acid Water Per Hour
BAWUL	Barrels acid water under load
B & B	Bell and bell
BB fraction	Butane-butene fraction
B/B	Back to Back, Barrels per barrel
BB	Bridged back
BBE	Bevel both ends
B.Bl	Base Blane
BBL	Barrel, barrels
B & CB	Beaded and centre beaded
BC	Barrels of condensate, bottom choke
BCF	Billion cubic feet
BCFD	Billion cubic feet per day
BCPD	Barrels condensate per day
BCPH	Barrels condensate per hour
BCPMM	Barrels condensate per million
BD	Barrels of distillate, budgeted depth
B/D	Barrels per day
B/dry	Bailed dry
bd	Board
BDA	Breakdown acid
Bd'A	Bois d'Arc
BDF	Broke (break) down formation
bd ft	Board foot; board feet
BD-MLW	Barge deck to mean low water
BDO	Barrels diesel oil
BDP	Breakdown pressure
BDPD	Barrels distillate per day
BDPH	Barrels distillate per hour
BDT	Blow-down test

B.E	Bevelled end
Be	Berea
Be	Baume
Bear R	Bear River
bec	Becoming
Beck	Beckwith
Bel	Beldon
Bel C	Belle City
Bel F	Belle Fourche
Belm	Belemnites
Ben	Benoist (Bethel) sand
Ben	Benton
Bent	Bentonite, bontinitic
bev	Bevel, beveled, as for welding
b & f	Ball and flange
bf	Buff
BF	Barrels fluid
BFO	Barrels frac oil
BFPD	Barrels fluid per day
BFPH	Barrels fluid per hour
BFW	Barrels formation water, boiler feed water
B/H	Barrels per hour
BHA	Bottom-hole assembly
BHC	Bottom-hole choke
BHFP	Bottom-hole flowing pressure
BHL	Bottom-hole location
BHM	Bottom-hole money
BHN	Brinell hardness number
B Hn	Big Horn
BHP	Bottom-hole pressure
bhp	Brake horsepower
bhp-hr	Brake horsepower hour
BHPC	Bottom-hole pressure, closed (See also SIBHP and BHSIP)
BHPF	Bottom-hole pressure, flowing
BHPS	Bottom-hole pressure survey
B/hr	Barrels per hour
BHSIP	Bottom-hole shut-in pressure
BHT	Bottom-hole temperature

Big.	Bigenerina
Big. f.	Bigenerina floridana
Big. h.	Bigenerina humblei
Big. nod.	Bigenerina nodosaria
B. Inj.	Big Injun
bio	Biotite
bit	Bitumen, bituminous
bkdn	Breakdown
bkr	Breaker
BL	Barrels load
B/L	Bill of lading
bl	Blue
BL&AW	Barrels load & acid water
bld	Bailed, blind (flange)
bldg	Building, bleeding, bleeding gas
bldg drk	Building derrick
bldg rds	Building roads
bldo	Bleeding oil
bldrs	Boulders
blg	Bailing
Blin	Blinebry
BL/JT	Blast joint
blk	Black, block
Blk Lf	Black Leaf
Blk Li	Black Lime
blk lnr	Blank liner
blnd	Blend/blended/blending
blndr	Blender
BLO	Barrels load oil
blo	Blow
BLOR	Barrels load oil recovered
Blos	Blossom
BLOYR	Barrels load oil yet to recover
blr	Bailer
B. Ls	Big Lime
blts	Bullets
BLW	Barrels load water
BM	Barrels mud, bench mark, Black Magic (mud)
B/M	Bill of material

BMEP	Brake mean effective pressure
BMI	Black malleable iron
bmpr	Bumper
bn	Brown
bnd	Band(ed)
bndry	Boundary
bnish	Brownish
BNO	Barrels new oil
bnz	Benzene
BO	Barrels oil, backed out (off)
BOCD	Barrels oil per calendar day
BOCS	Basal Oil Creek sand
BOD	Barrels oil per day
Bod	Bodcaw
BOE	Bevel one end
BOE	Blow out equipment
Bol.	Bolivarensis
Bol. a.	Bolivina a.
Bol. flor.	Bolivina floridana
Bol. p.	Bolivina perca
Bonne	Bonneterre
BOP	Blowout preventer
BOPCD	Barrels oil per calendar day
BOPD	Barrels oil per day
BOPH	Barrels oil per hour
BOPPD	Barrels oil per producing day
BOS	Brown oil stain
bot	Bottom
BP	Back pressure, boiling point, bridge plug, bull plug, bulk plant
BP	Bearpaw, Base Pennsylvanian
BPCD	Barrels per calendar day
BPH	Barrels per hour
BPLO	Barrels of pipeline oil
BPLOPD	Barrels of pipeline oil per day
BPM	Barrels per minute
BP Mix	Butane and propane miz
BPV	Back pressure valve
BPWPD	Barrels per well per day
BR	Building rig, building road

brach	Brachiopod
brec	Breccia
brg	Bearing
Brid	Bridger
brit	Brittle
B. Riv	Black River
brk	Break (broke)
brkn	Broken
brkn sd	Broken sand
brksh	Brackish (water)
brkt(s)	Bracket(s)
brn or br	Brown
Brn Li	Brown lime
brn sh	Brown shale
Brom	Bromide
brtl	Brittle
bry	Bryozoa
B & S	Bell and spigot
BS	Basic sediment, bottom sediment, bottom settlings, Bone Spring
BS&W	Basic sediment & water
B/S	Bill of sale, base salt
B/SD	Barrels per stream day (refinery)
BSE	Bevel small end
BSFC	Brake specific fuel consumption
bsg	Bushing
B&S ga	Brown and Sharpe gauge
bskt	Basket
B slt	Base of the salt
bsmt	Basement
BSPL	Base plate
BSUW	Black sulfur water
BSW	Barrels salt water
BSWPD	Barrels salt water per day
BSWPH	Barrels salt water per hour
BT	Benoist (Bethel) sand
BTDC	Before top dead center
btm	Bottom
btm chk	Bottom choke
btmd	Bottomed

btry	Battery
BTU	British thermal unit
btw	Between
BTX (unit)	Benzene, toluene, xylene (Unit)
Buck	Buckner
Buckr	Buckrange
Bul. text.	Buliminella textularia
Bull W	Bullwaggon
bunr	Burner
Burg	Burgess
butt	Buttress thread
BV/WLD	Beveled for welding
BW	Barrels of water, boiled water, butt weld
BW/D	Barrels of water per day
BW ga	Birmingham (or Stubbs) iron wire gauge
BWL	Barrels water load
BWOL	Barrels water over load
BWPD	Barrels of water per day
BWPH	Barrels of water per hour
bx	Box(es)

C

C	Center (land description), Centigrade temp. scale
C/	Contractor (i.e., C/John Doe)
c	Coarse(ly)
C/A	Commission agent
C&A	Compression and absorption plant
C to C	Center to center
C to E	Center to end
C to F	Center to face
Cadd	Caddell
CAG	Cut across grain
cal	Caliper survey, calorie, calcite, calcitic, caliche

Calc	Calcium, calcareous, calcerenite, calculate(d)
calc OF, COF	Calculated open flow (potential)
calc gr	Calcium base grease
calc	Calceneous
Calv	Calvin
Cam.	Camerina
Camb	Cambrian
Cane R	Cane River
Cany	Canyon
Cany Crk	Canyon Creek
CAOF	Calculated absolute open flow
cap	Capacity, capacitor
Cap	Capitan
Car	Carlile
carb	Carbonaceous
carb tet	Carbon tetrachloride
Carm	Carmel
Casp	Casper
cat	Catalyst, catalytic, catalog
CAT	Carburetor air temperature, catalog, catalyst, catalytic
Cat	Catahoula
Cat ckr	Catalytic cracker
Cat Crk	Cat Creek
cath	Cathodic
caus	Caustic
cav	Cavity
CB	Counterbalance (pumping equip.), core barrel, changed(ing) bits
cc	Cubic centimeter
CC	Carbon copy, casing cemented (depth), closed cup
C & C	Circulating and conditioning
C-Cal	Contact caliper
CCHF	Center of casinghead flange
Cck	Casing choke
CCL	Casing collar locator
CCLGO	Cat cracked light gas oil
CCM	Condensate-cut mud

CCP	Critical compression pressure
CCPR	Casing collar perforating record
CCR	Conradson carbon residue
CCR	Critical compression ratio
CCU	Catalytic Cracking Unit
ccw	Counterclockwise
CD	Contract depth, calendar day
CDM	Continuous dipmeter survey
Cdr Mtn	Cedar Mountain
cdsr	Condensor
Cdy	Cody (Wyoming)
cell	Cellar, cellular
cem	Cement(ed)
CEMF	Counter electromotive force
Ceno	Cenozoic
cent	Centralizers
centr	Centrifugal
ceph	Cephalopod
Cert	Ceratobulimina eximia
CF	Casing flange
Cf	Cockfield
CF	Cubic feet, clay filled
CFBO	Companion flanges bolted on
CFG	Cubic feet gas
CFGPD	Cubic feet gas per day
CFGH	Cubic feet of gas per hour
CFM	Cubic feet per minute
CFOE	Companion flange one end
CFR	Cement friction reducer
CFS	Cubic feet per second
c-gr	Coarse-grained
CG	Corrected gravity, center of gravity
cg	Coring
cglt	Conglomerate, conglomeritic
cgs	Centimeter-gram-second system
CH	Casinghead (gas)
C/H	Cased hole
ch	Chert, choke
chal	Chalcedony
Chapp	Chappel

Char	Charles
Chatt	Chattanooga shale
chem	Chemical, chemist, chemistry
chem prod	Chemical products
Cher	Cherokee
Ches	Chester
CHF	Casinghead flange
CHG	Casinghead gas
chg	Charge, charged, charging
chng	Change, changed, changing
Chim H	Chimney Hill
Chim R	Chimney Rock
Chin	Chinle
chit	Chitin(ous)
chk	Choke, chalk
Chkbd	Checkerboard
chkd	Checked
chky	Chalky
chl	Chloride(s), chloritic
chl log	Chlorine log
Chou	Chouteau lime
CHP	Casinghead pressure
chrm	Chairman
chrome	Chromium
chromat	Chromatograph
cht	Chart, chert
chty	Cherty
Chug	Chugwater
CI	Cast-iron, contour interval (map)
Cib	Cibicides
Cib h	Cibicides hazzardi
CIBP	Cast-iron bridge plug
CI engine	Compression-ignition engine
C.I.F.	Cost, insurance and freight
Cima	Cimarron
CIP	Cement in place, closed-in pressure
cir	Circle, circular, circuit
circ	Circulate, circulating, circulation
cir mils	Circular mils
Cis	Cisco

ck	Check, cake
cksn	Chicksan
Ck Mt	Cook Mountain
CL	Carload
C/L	Center line
Clag	Clagget
Claib	Claiborne
Clarks	Clarksville
clas	Clastic
Clav	Clavalinoides
Clay	Clayton, Claytonville
Cleve	Cleveland
Clfk	Clearfork
Cliff H	Cliff House
CLMP	Canvas-lined metal petal basket
cin (d) (g)	Clean, cleaned, cleaning
Clov	Cloverly
clr	Clear, clearance
clrg	Clearing
clsd	Closed
clyst	Claystone
cm	Centimeter
CMC	Sodium carboxymethylcellulose
Cmchn	Comanchean
Cmpt	Compact
cm/sec	Centimeters per second
cmt (d) (g)	Cement (ed) (ing)
cmtr	Cementer
CN	Cetane number
cncn	Concentric
cntf	Centrifuge
cntl	Control(s)
cntr	Center(ed), controller, container
Cnty	County
cnvr	Conveyor
CO	Clean out, cleaning out, cleaned out, crude oil, circulated out
c/o	Care of
COBOL	Common Business Oriented Language
COC	Cleveland open cup

Coco	Coconino
Cod	Codell
coef	Coefficient
COF	Calculated open flow
COG	Coke oven gas
COH	Coming out of hole
COL	Colored, column
Col ASTM	Color, American Standard Test Method
Cole J	Cole Junction
Col Jct	Coleman Junction
coll	Collect, collected, collection, collecting
colr	Collar
Com	Comanche
Com	Comatula
com	Common
comb	Combined, combination
coml	Commercial
comm	Community, communitized, communication
comm	Commission, commenced
commr	Commissioner
comp	Complete, completed, completion
Com Pk	Comanche Peak
comp nat	completed natural
compnts	Components
compr	Compressor
compr sta	Compressor station
compt	Compartment
con	Consolidated
conc	Concentric, concentrate, concrete
conc	Concretion(ary)
conch	Conchoidal
concl	Conclusion
cond	Condensate, conditioned, conditioning,
condr	Conductor (pipe)
condt	Conductivity
conf	Confirm, confirmed, confirming

confl	Conflict
cong	Conglomerate(itic)
conn	Connection
cono	Conodonts
consol	Consolidated
const	Constant, construction
consv	Conserve, conservation
cont(d)	Continue, continued
contam	Contaminated, contamination
contr	Contractor
contr resp	Contractor's responsibility
contrib	Contribution
conv	Converse
Co. Op.	Company-operated
co-op	Cooperative
Co. Op. S.S.	Company-operated service stations
coord	Coordinate
COP	Crude oil purchasing
coq	Coquina
cor	Corner
Corp	Corporation
corr	Correct (ed) (ion), corrosion, corrugated
correl	Correlation
corres	Correspondence
CO & S	Clean out & shoot
COTD	Cleaned out to total depth
Cott G	Cottage Grove
Counc G	Council Grove
CP	Casing point, casing pressure
CP	Chemically pure
C & P	Cellar & pits
cp	Centipoise
CPA	Certified public accountant
CPC	Casing pressure—closed
Cp Colo	Camp Colorado
CPF	Casing pressure—flowing
CPG	Cost per gallon or cents per gallon
cplg	Coupling
CPM	Cycles per minute

CPS	Cycles per second
CPSI	Casing pressure shut in
CR	Cold rolled, compression ratio
CR	Cow Run
CR	Cane River
cr (d), (g), (h)	Core, cored, coring, core hole
CRA	Chemically retarded acid
crbd	Crossbedded
CRC	Coordinating Research Council, Inc.
CR Con	Carbon Residue (Conradson)
cren	Crenulated
Cret	Cretaceous
Crin	Crinoid(al)
Cris	Cristellaria
crit	Critical
crkg	Cracking
Crkr	Cracker
cr moly	Chrome molybdenum
crn blk	Crown block
crnk	Crinkled
Crom	Cromwell
crs	Coarse(ly)
crypto-xln	Cryptocrystalline
cryst	Crystalline
CS	Cast steel, carbon steel
CS	Casing seat
cs	Centistokes
CSA	Casing set at
cse gr	Coarse grained
csg	Casing
csg hd	Casing head
csg press	Casing pressure
csg pt	Casing point
CSL	County school lands, center section line
CT	Cable tools
CTC	Consumer tank car
ctd	Coated
CTD	Corrected total depth
ctg(s)	Cuttings

CTHF	Center of tubing flange
Ctlmn	Cattleman
ctn	Carton
Ctnwd	Cottonwood
CTP	Cleaning to pits
ctr	Center
CTT	Consumer transport truck
ctw	Coated and wrapped
CTW	Consumer tank wagon
cu	Cubic
CU	Clean up
cu ft	Cubic foot
cu ft/bbl	Cubic feet per barrel
cu ft/min	Cubic feet per minute
cu ft/sec	Cubic feet per second
cu in	Cubic inch
culv	Culvert
cum	Cumulative
cu m	Cubic meter
Cur	Curtis
cush	Cushion
Cut B	Cut Bank
cutbk	Cutback
Cutl	Cutler
Cut Oil	Cutting oil
Cut Oil Act Sul-Dk	Cutting oil-active sulphurized-dark
Cut Oil Act Sul-Trans	Cutting oil-active-sulphurized transparent
Cut Oil Inact Sul	Cutting oil-inactive-sulphurized
Cut Oil Sol	Cutting oil soluble
Cut Oil St Mrl	Cutting oil-straight mineral
cu yd	Cubic yard
CV	Cotton Valley, control valve
cvg(s)	Caving(s)
CW	Continuous weld
cw	Clockwise
C/W	Complete with
C & W	Coat and wrap (pipe)
CWP	Cold working pressure

cwt	Hundred weight
CX	Crossover
Cyc	Cyclamina
Cycl canc	Cyclamina cancellata
cyl	Cylinder
cyl stk	Cylinder stock
cyp	Cypridopsis
Cy Sd	Cypress Sand
Cz	Carrizo

D

d-1-s	Dressed one side
d-2-s	Dressed two sides
d-4-s	Dressed four sides
D-2	Diesel No. 2
DA	Daily allowable
D & A	Dry and abandoned
DAIB	Daily average injection, barrels
Dak	Dakota
Dan	Dantzler
Dar	Darwin
DAR	Discovery allowable requested
dat	Datum
db	Decibel
DB	Drilling break
D & B	Dun & Bradstreet
d/b/a	Doing business as
DBO	Dark brown oil
DBOS	Dark brown oil stains
DC	Delayed coker, direct current, drill collar, dually complete(d), development well—carbon dioxide, diamond core
D & C	Drill and complete
DCB	Diamond core bit
DCLSP	Digging slush pits, digging cellar, or digging cellar and slush pits

DCM	Distillate-cut mud
DCS	Drill collars
D/D	Day to day
dd	Dead
DD	Degree day, drilling deeper
d-d-l-s-l-e	Dressed dimension one side and one edge
d-d-4-s	Dressed dimension four sides
D & D	Desk and Derrick
DDD	Dry desiccant dehydrator
DDT	Dichloro-diphenyl-trichloroethane
DE	Double end
Deadw	Deadwood
deaer	Deaerator
deasph	Deasphalting
debutzr	Debutanizer
dec	Decimal
decr	Decrease (d) (ing)
deethzr	Deethanizer
defl	Deflection
deg	Degonia
deisobut	Deisobutanizer
Dela	Delaware
Del R	Del Rio
delv	Delivery, delivered, deliverability
delv pt	Delivery point
demur	Demurrage
dend	Dendrite(ic)
DENL	Density log
dep	Depreciation
depl	Depletion
deprec	Depreciation
deprop	Depropanizer
dept	Department
desalt	Desalter
desc	Description
Des Crk	Desert Creek
Des M	Des Moines
desorb	Desorbent
desulf	Desulferizer

det	Details(s), detector
deterg	Detergent
detr	Detrital
dev	Deviate, deviation
Dev	Devonian
devel	Develop (ed) (ment)
dewax	Dewaxing
Dext	Dexter
DF	Derrick floor, diesel fuel
DFE	Derrick floor elevation
DFO	Datum faulted out
DFP	Date of first production
DG	Development gas well, draft gage, dry gas
DH	Development well–helium
DHC	Dry hole contribution
DHDD	Dry hole drilled deeper
DHM	Dry hole money
DHR	Dry hole reentered
dia	Diameter
diag	Diagram, diagonal
diaph	Diaphragm
dichlor	Dichloride
diethy	Diethylene
diff	Differential, difference, different
dilut	Diluted
dim	Dimension, diminish, diminishing
Din	Dinwoody
dir	Direct, direction, director
dir sur	Directional survey
Disc	Discorbis
disc	Discount, discover (y) (ed) (ing)
Disc grav	Discorbis gravelli
disch	Discharge
Disc norm	Discorbis normada
Disc y	Discorbis yeguaensis
dism	Disseminated
disman	Dismantle
displ	Displaced, displacement
dist	Distance, distillate, distillation, district

distr	Distribute (d) (ing) (ion)
div	Division
dk	Dark
Dk Crk	Duck Creek
D/L	Density log
dlr	Dealer
DM	Datum, demand meter, dip meter, drilling mud
dml	Demolition
dmpr	Damper
dn	Down
dns	Dense
DO	Drill (ed) (ing) out, development oil well
D.O.	Division order
D/O	Division Office
do	Ditto
DOC	Diesel oil cement, drilled out cement
Doc	Dockum
doc	Document
doc-tr	Doctor-treating
DOD	Drilled out depth
dolo	Dolomite(ic)
dolst	Dolstone
dom	Domestic
dom AL	Domestic airline
DOP	Drilled out plug
Dorn H	Dornick Hills
Doth	Dothan
Doug	Douglas
doz	Dozen
DP	Data processing, dew point, drill pipe
DP	Double pole (switch)
D/P	Drill (ed) (ing) plug
DPDB	Double pole double base (switch)
DPDT	Double pole double throw (switch)
dpg	Deepening
DPM	Drill pipe measurement
dpn	Deepen
DPSB	Double pole single base (switch)
DPST	Double pole single throw (switch)

DPT	Deep pool test
dpt	Depth
dpt rec	Depth recorder
DPU	Drill pipe unloaded
dr	Drain, drive, drum, druse
DR	Development redrill (sidetrack)
Dr Crk	Dry Creek
drk	Derrick
DRL	Double random lengths
drl	Drill
drld	Drilled
drlg	Drilling
drlr	Driller
drng	Drainage
drpd	Dropped
drsy	Drusy
dry	Drier, drying
DS	Directional survey
DS	Drill stem
ds	dense
dsgn	Design
DSI	Drilling suspended indefinitely
dsl	Diesel (oil)
dsmtl(g)	Dismantle(ing)
DSO	Dead oil show
DSS	Days since spudded
DST	Drill stem test
dstl	Distillate
dstn	Destination
DSU	Development well—sulphur
DT	Drilling time
D/T	Driller's tops
DTD	Drillers total depth
dtr	Detrital
DTW	Dealer tank wagon
Dup	Duperow
dup	Duplicate
Dutch	Dutcher
DV	Differential valve (cementing)
DWA	Drilling with air

dwg	Drawing
dwks	Drawworks
DWM	Drilling with mud
dwn	Down
DWO	Drilling with oil
DWP	Dual (double) wall packer
DWSW	Drilling with salt water
DWT	Dead weight tester
DX	Development well workover
dx	Duplex
dyn	Dynamic

E

E	East
E/2, E/4	East half, quarter, etc.
ea	Each
EAM	Electric accounting machines
Earls	Earlsboro
Eau Clr	Eau Claire
E/BL	East boundary line
ecc	Eccentric
Ech	Echinoid
ECM	East Cimarron Meridian (Oklahoma)
Econ	Economics, economy, economizer
Ect	Ector (county, Tex.)
Ed lm	Edwards lime
EDP	Electronic data processing
Educ	Education
Edw	Edwards
E/E	End to end
EF	Eagle Ford
eff	Effective, efficiency
effl	Effluent
EFV	Equilibrium flash vaporization
e.g.	For example
Egl	Eagle

Eglwd	Englewood
EHP	Effective horsepower
eject	Ejector
E/L	East Line
Elb	Elbert
elec	Electric(al)
elem	Element(ary)
elev	Elevation, elevator
Elg	Elgin
el gr	Elevation ground
ell(s)	Elbow(s)
Ellen	Ellenburger
Elm	Elmont
EL/T	Electric log tops
EM	Eagle Mills
Emb	Embar
EMS	Ellis-Madison contact
emer	Emergency
EMF	Electromotive force
empl	Employee
emul	Emulsion
encl	Enclosure
End	Endicott
endo	Endothyra
eng	Engine
engr(g)	Engineer(ing)
enl	Enlarged
enml	Enamel
Ent	Entrada
E/O	East offset
EO	Emergency order
Eoc	Eocene
EOF	End of file
EOL	End of line
EOM	End of month
EOQ	End of quarter
EOR	East of Rockies
EOY	End of year
EP	End point, extreme pressure
Epon	Eponides

Ep y	Eponides yeguaensis
eq	Equal, equalizer, equation
equip	Equipment
equiv	Equivalent
erect	Erection
Eric	Ericson
ERW	Electric resistance weld
est	Estate, estimate (d) (ing)
ETA	Estimated time of arrival
et al.	And others
et con	And husband
eth	Ethane
ethyle	Ethylene
et seq	And the following
et ux	And wife
et vir	And husband
Eu	Eutaw
EUE	External upset end
euhed	Euhedral
ev	Electron volts
eval	Evaluate
evap	Evaporation, evaporite
ev-sort	Even-sorted
EW	Electric weld, exploratory well
E of W/L	East of west line
Ex	Exter
exc	Excavation
exch	Exchanger
excl	Excellent
exh	Exhaust, exhibit
exist	Existing
exp	Expansion, expense
expl	Exploratory, exploration
exp plg	Expendable plug
explos	Explosive
expir	Expiration, expire, expired, expiring
exr	Executor
Exrx	Executrix
exst	Existing
ext(n)	Extended, extension

ext	External
Ext M/H	Extension manhole
extr	Exterior
extrac	Extraction
EYC	Estimated yearly consumption

F

°F.	Degree Fahrenheit
F/	Flowed, flowing
fab	Fabricate(d)
FAB	Faint air blow
fac	Facet(ed)
FACO	Field authorized to commence operations
fail	Failure
Fall Riv	Fall River
Farm	Farmington
FAO	Finish all over
FARO	Flowed(ing) at rate of
fau	Fauna
FB	Fresh break
FBH	Flowing by heads
FBHP	Flowing bottom hole pressure
FBHPF	Final bottom hole pressure flowing
FBHPSI	Final bottom hole pressure shut-in
FBP	Final boiling point
FC	Filter cake, float collar
FC	Fixed carbon
FCC	Fluid catalytic cracking
FCP	Flowing casing pressure
FCV	Flow control valve
FD	Feed, floor drain
F-D	Formation density
F & D	Flanged and dished (heads), faced and drilled
fdn	Foundation

FDL	Formation density log
fdr	Feeder
fed	Federal
FE/L	From east line
FELA	Federal Employers Liability Act
Ferg	Ferguson
ferr	Ferruginous
fert	Fertilizer
Fe-st	Ironstone
FF	Flat face
FF	Frac finder (Log), full of fluid, fishing for
F & F	Fuels & fractionation
F to F	Face to face
FFA	Female to female angle
FFG	Female to female globe (valve)
FFO	Furnace Fuel Oil
FFP	Final flowing pressure
f-gr	Fine-grained
F.G.	Fracture gradient
FGIH	Finish going in hole
FGIW	Finish going in with—
F/GOR	Formation gas-oil ratio
FH	Full hole
FHP	Final hydrostatic pressure
FI	Flow indicator
fib	Fibrous
FIC	Flow indicating controller
fig	Figure
FIH	Fluid in hole
fill	Fillister
filt	Filtrate
fin	Final, finish, finished
fin drlg	Finished drilling
FIRC	Flow indicating ratio controller
fis	Fissure
fish	Fishing
fisl	Fissile
FIT	Formation interval tester
fix	Fixture

FJ	Flush joint
FL	Floor, fluid level, flow line, flush
fl/	Flowed or flowing
fl	Fluid
F & L	Fuels & Lubricants
FLA	Ferry Lake anhydrite
flat	Flattened
Flath	Flathead
Fl–COC	Flash Point, Cleveland Open Cup
fld	Failed, feldspar (thic), field
flex	Flexible
flg (d) (s)	Flange (d) (s), flowing
Flip	Flippen
flk	Flaky
flo	Flow
Flor Fl	Florence Flint
flt	Float, fault
fltg	Floating
flu	Flue, fluid
fluor	Fluorescence, fluorescent
flshd	Flushed
flw (d) (g)	Flowed, flowing
Flwg Pr.	Flowing pressure
Flwrpt	Flowerpot
fm	Formation
FM	Frequency meter, frequency modulation
f'man	Foreman
Fm W	Formation water
fn	fine
FNEL	From northeast line
FNL	From north line
fnly	Finely
fnt	Faint
FNWL	From northwest line
FO	Farmout, fuel oil, full opening, faulted out
FOB	Free on board
FOCL	Focused log
F.O.E.	Fuel oil equivalent

FOE—WOE	Flanged one end, welded one end
FOH	Full open head
	(grease drum 120 lb)
fol	Foliated
Forak	Foraker
foram	Foraminifera
Fort	Fortura
foss	Fossiliferous
FOT	Flowing on test
Fount	Fountain
Fox H	Fox Hills
FP	Final pressure, flowing pressure,
	freezing point
FPI	Free point indicator
fpm	Feet per minute
FPO	Field purchase order
fprf	Fireproof
fps	Feet per second
fps	Foot-pound-second (system)
FPT	Female pipe thread
FQG	Frosted quartz grains
FR	Flow recorder, flow rate, feed rate
fr	Fair, fractional, frosted, from, front
FRA	Friction reducing agent
frac (d) (s)	Fracture, fractured, fractures
fract	Fractionation, fractionator
frag	Fragment
fran	Franchise
Franc	Franconia
FRC	Flow recorder control
Fred	Fredericksburg, Fredonia
fr E/L	From east line
freq	Frequency
Frgy	Froggy
fri	Friable
fr N/L	From north line
F-R Oil	Fire-resistant oil
Fron	Frontier
fros	Frosted
FRP	Fiberglass reinforced plastic

FRR	Final report for rig
fr S/L	From south line
frs	Fresh
frt	Freight
Fruit	Fruitland
FRW	Final report for well
frwk	Framework
fr W/L	From west line
frzr	Freezer
FS	Feedstock, forged steel, float shoe
F/S	Front & side, flange x screwed
F & S	Flanged and spigot
FSEL	From southeast line
fsg	Fishing
FSIP	Final shut-in pressure
FSL	From south line
FSP	Flowing surface pressure
FST	Forged steel
FSWL	From southwest line
FS&WLs	From south and west lines
ft	Foot, feet
Ft C	Fort Chadborne
ft-c	Foot-candle
ftg	Fittings, footing, footage
Ft H	Fort Hayes
ft/hr	Feet per hour
ft lb	Foot-pound
ft lbs/hr	Foot-pounds per hour
ft/min	Feet per minute
FTP	final tubing pressure, flowing tubing pressure
Ft R	Fort Riley
FTS	Fluid to surface
ft/sec	Feet per second
Ft U	Fort Union
Ft W	Fort Worth
FU	Fill up
Full	Fullerton
furf	Furfural
furn	Furnace

Furn & fix	Furniture and fixtures
Fus	Fuson
Fussel	Fusselman
Fusul	Fusulinid
fut	Future
FV	Funnel viscosity
fvst	Favosites
FW	Fresh water
FWC	Field wildcat
fwd	Forward
FWD	Four-wheel drive
FWL	From west line
fxd	Fixed
f/xln	Finely-crystalline
FYI	For your information

G

G	Gas
g	Gram
GA	Gallons acid
ga	Gage (d) (ing)
GAF	Gross acre feet
gal	Gallon, gallons
Gall	Gallatin
gal/Mcf	Gallons per thousand cubic feet
gal/min	Gallons per minute
gal sol	Gallons of solution
galv	Galvanized
gaso	Gasoline
gast	Gastropod
GB	Gun barrel
GBDA	Gallons breakdown acid
GC	Gas-cut
g-cal	Gram-calorie
GCAW	Gas-cut acid water
GCD	Gas-cut distillate

GCLO	Gas-cut load oil
GCLW	Gas-cut load water
GCM	Gas-cut mud
GCO	Gas-cut oil
GCPD	Gallons condensate per day
GCPH	Gallons condensate per hour
GCR	Gas-condensate ratio
GCSW	Gas-cut salt water
GCW	Gas-cut water
gd	Good
GD	Glen Dean lime
Gdld	Goodland
gd o&t	Good odor & taste
GDR	Gas-distillate ratio
Gdwn	Goodwin
GE	Grooved ends
G egg	Goose egg
gel	Jelly-like colloidal suspension
gen	Generator
genl	General
Geo	Georgetown
Geol	Geologist, geological, geology
GFLU	Good fluorescence
GGW	Gallons gelled water
GHO	Gallons heavy oil
Geop	Geophysics, geophysical
GH	Greenhorn
GI	Gas injection
Gib	Gibson
GIH	Going in hole
gil	Gilsonite
Gilc	Gilcrease
GIW	Gas-injection well
GJ	Ground joint
GL	Gas lift, ground level
G/L	Gathering line
gl	Glassy
glau	Glauconitic, glauconite
Glen	Glenwood
Glna	Galena

GLO	General Land Office (Texas)
Glob	Globigerina
Glor	Glorieta
GLR	Gas-liquid ratio
gls	Glass
glyc	Glycol
gm	Gram
GM	Ground measurement (elevation)
G.M.	Gravity meter
GMA	Gallons mud acid
gm-cal	Gram calorie
G&MCO	Gas & mud-cut oil
g mole	gram molecular weight
gmy	Gummy
gnd	grained (as in fine-grained)
gns	Gneiss
GO	Gas odor, gallons oil
G&O	Gas and oil
GOC	Gas-oil contract
G&OCM	Gas and oil-cut mud
GODT	Gas odor distillate taste
Gol	Golconda lime
Good L	Goodland
GOPH	Gallons of oil per hour
GOPD	Gallons of oil per day
GOR	Gas-oil ratio
Gor	Gorham
Gouldb	Gouldbusk
gov	Governor
govt	Government
GP	Gas pay, gasoline plant
G/P	Gun perforate
GPC	Gas purchase contract
GPD	Gallons per day
GPG	Grains per gallon
GPH	Gallons per hour
GPM	Gallons per minute
gpm	Gallons per thousand cubic feet
GPS	Gallons per second
GR	Gamma ray, Glen Rose

gr	Ground, grade, grain, grease
GRA	Gallons regular acid
grad	Gradual, gradually
gran	Granite, granular
Granos	Graneros
Gran W	Granite Wash
grap	Graptolite
gr API	Gravity, °API
grav	Gravity
Gray	Grayson
Grayb	Grayburg
grd	Ground
grdg	Grading
GRDL	Guard log
grd loc	Grading location
G.Riv	Gull River
G Rk	Gas Rock
Grn	Green
grnlr	granular
Grn Riv	Green River
Grn sh	Green shale
gr roy	Gross royalty
grs	Gross
Gr Sd	Gray sand
grt	Grant (of land)
grtg	Grating
grty	Gritty
grv	Grooved
grvt	Gravitometer
gr wt	Gross weight
gry	Gray
GS	Gas show
GSC	Gas sales contract
GSG	Good show of gas
GSI	Gas well shut-in
gskt	Gasket
GSO	Good show of oil
GSW	Gallons salt water
gsy	Greasy
GTS	Gas to surface (time)

GTSTM	Gas too small to measure
GTY	Gravity
GU	Gas Unit
Guns	Gunsite
GV	Gas volume
gvl	Gravel
GVLPK	Gravel packed
GVNM	Gas volume not measured
GW	Gallons water, gas well
GWC	Gas-water contact
GWG	Gas-well gas
CWPH	Gallons of water per hour
gyp	Gypsum
Gyp Sprgs	Gypsum Springs
gypy	Gypsiferous
Gyr	Gyroidina
Gyr sc	Gyroidina scal
gywk	Graywacke

H

Hackb	Hackberry
Hara	Haragan
Hask	Haskell
Haynes	Haynesville
haz	Hazardous
HB	Housebrand (regular grade of gasoline)
Hberg	Hardinsburg sand (local)
HBP	Held by production
hbr	Harbor
HC	Hydrocarbon
HCO	Heavy cycle oil
HCV	Hand-control valve
HD	High detergent, heavy duty, Hydril
hd	Hard, head
hd li	Hard lime

hdns	Hardness
hdr	Header
hd sd	Hard sand
hdl	Handle
hdwe	Hardware
Heeb	Heebner
hem	Hematite
Her	Herington
Herm	Hermosa
het	Heterostegina
HEX	Heat exchanger
hex	Hexagon(al), hexane
hfg	Hydrofining
HFO	Heavy fuel oil, hole full of oil
HF Sul W	Hole full of sulphur water
HFSW	Hole full of salt water
HFW	Hole full of water
HGCM	Heavily gas-cut mud
HGCW	Heavily gas-cut water
HGOR	high gas-oil ratio
hgr	Hanger
hgt	Height
HH	Hand hole, hydrostatic head
H H P	Hyraulic horsepower
Hick	Hickory
Hill	Hilliard
hily	Highly
hky	Hackly
HLSD	High-level shut-down
HO	Heavy oil, heating oil
hock	Hockleyensis
HOCM	Heavily oil-cut mud
HOCW	Heavily oil-cut water
Hog	Hogshooter
HO&GCM	Heavily oil-and-gas-cut mud
Holl	Hollandberg
Home Cr	Home Creek
hop	Hopper
horiz	Horizontal
Hosp	Hospah

Hov	Hoover
Hox	Hoxbar
HP	High pressure, horsepower, hydraulic pump, hydrostatic pressure
HPF	Holes per foot
HPG	High-pressure gas
hp hr	Horsepower hour
HQ	Headquarters
HRS	Hot-rolled steel
hr	Hour, hours
HRD	High-resolution dipmeter
hrs	Heirs
HSD	Heavy steel drum
ht	Heater treater, heat treated, high temperature, high tension
htr	Heater
HTSD	High-temperature shut-down
Humb	Humblei
Hump	Humphreys
Hun	Hunton
HV	High viscosity
HVI	High viscosity index
hvly	Heavily
hvy	Heavy
HWCM	Heavily water-cut mud
HWP	Hookwall packer
hwy	Highway
HX	Heat exchanger
HYD	Hydril thread, hydraulic
HYDA	Hydril Type A joint
HYDCA	Hydril Type CA joint
HYDCS	Hydril Type CS joint
hydtr	Hydrotreater
Hyg	Hygiene
Hz	Hertz (new name for electrical cycles per second)

I

IAB	Initial air blow
IB	Impression block
IB	Iron body (valve)
IBBC	Iron body, brass core (valve)
IBBM	Iron body, brass (bronze) mounted (valve)
IBHP	Initial bottom-hole pressure
IBHPF	Initial bottom-hole pressure flowing
IBHPSI	Initial bottom-hole pressure shut-in
IBP	Initial boiling point
IC	Iron case
ID	Inside diameter
I.D. Sign	Identification sign
Idio	Idiomorpha
IF	Internal flush
IFP	Initial flowing pressure
Ign	Igneous
IGOR	Injection gas-oil ratio
IHP	Initial hydrostatic pressure
IHP	Indicated horsepower
IHPHR	Indicated horsepower hour
IJ	Integral joint
imbd	Imbedded
immed	Immediate(ly)
Imp	Imperial
imperv	Impervious
Imp gal	Imperial gallon
IMW	Initial mud weight
in	Inch(es)
inbd	Interbedded
inbdd	Inbedded
Inc	Incorporated
incd	Incandescent
incin	Incinerator
incl	Include, included, including

incls	Inclusions
incolr	Intercooler
incr	Increase (d) (ing)
ind	Induction
indic	Indicate, indication, indicates
indiv	Individual
indr	Indurated
indst	Indistinct
Inf. L	Inflammable liquid
info	Information
Inf. S	Inflammable solid
ingr	Intergranular
in. Hg	Inches mercury
inhib	Inhibitor
init	Initial
inj	Injection, injected
Inj Pr	Injection pressure
inl	Inland, inlet
inlam	Interlaminated
in-lb	Inch-pound
Inoc	Inoceramus
INPE	Installing, installed, pumping equipment
ins	Insurance, insulate, insulation
in/sec	Inches per second
insp	Inspect, inspected, inspecting, inspection
inst	Install (ed) (ing), instantaneous, institute
instl	Installation(s)
instr	Instrument, instrumentation
insul	Insulate
int	Interest, interval, internal, interior
interbd	Interbedded
inter-gran	Intergranular
inter-lam	Interlaminated
inter-xln	Inter-crystalline
intgr	Integrator
intl	Interstitial
intr	Instrusion
ints	Intersect
intv	Interval

inv	Invert, inverted, invoice
inven	Inventory
invrtb	Invertebrate
I/O	Input/output
IP	Initial potential, initial production, initial pressure
IPA	Isopropyl alcohol
IPE	Install(ing) pumping equipment
IPF	Initial potential flowed, initial production flowed(ing)
IPG	Initial production gas lift
IPI	Initial production on intermitter
IPL	Initial production plunger lift
IPP	Initial production pumping
IPS	Initial production swabbing
IPS	Iron pipe size
IPT	Iron pipe thread
IR	Injection rate
Ire	Ireton
irreg	Irregular
irid	Iridescent
IRS	Internal Revenue Service
irst	Ironstone
IS	Inside screw (valve)
ISIP	Initial shut-in pressure (DST), instantaneous shut-in pressure (frac)
isom	Isometric
isoth	Isothermal
ITD	Intention to drill
IUE	Internal upset ends
Ives	Iverson
IVP	Initial vapor pressure
IW	Injection well

J

J&A	Junked and abandoned
jac	Jacket
Jack	Jackson
Jasp	Jasper(oid)
Jax	Jackson sand
JB	Junk basket, junction box
jbr	Jobber
JC	Job complete
jct	Junction
jdn	Jordan
Jeff	Jefferson
JINO	Joint interest non-operated (property)
jmd	Jammed
jnk	Junk(ed)
J/O	Joint operation
JOA	Joint operating agreement
JOP	Joint operating provisions
JP	Jet perforated
JP/ft	Jet perforations per foot
JP fuel	Jet propulsion fuel
JSPF	Jet shots per foot
jt(s)	Joint(s)
Jud R	Judith River
Jur	Jurassic
juris	Jurisdiction
JV	Joint venture
Jxn	Jackson

K

K	Kelvin (temperature scale)
Kai	Kaibab
kao	Kaolin

Kay	Kayenta
KB	Kelly bushing
KBM	Kelly bushing measurement
KC	Kansas City
kc	Kilocycle
kcal	Kilocalorie
KD	Kiln dried, kincald lime
KDB	Kelly drive bushing
KDBE	Kelly drive bushing elevation
KDB-LDG FLG	Kelly drill bushing to landing flange
KDB-MLW	Kelly drill bushing to mean low water
KDB-Plat	Kelly drill bushing to platform
Ke	Keener
Keo-Bur	Keokuk–Burlington
kero	Kerosine
ket	Ketone
Key	Keystone
kg	Kilogram
kg-cal	Kilogram calorie
kg-m	Kilogram-meter
Khk	Kinderhook
KHz	Kilohertz (see Hz–Hertz)
Kia	Kiamichi
Kib	Kibbey
Kin	Kinematic
Kin	Kincaid lime
kip	One thousand pounds
kip-ft	One thousand foot-pounds
Kirt	Kirtland
kld	Killed
km	Kilometer
KMA	KMA sand
KO	Kicked off, knock out
Koot	Kootenai
KOP	Kickoff point
Kri	Krider
KV	Kinematic viscosity
kv	Kilovolt

kva	Kilovolt-ampere
kvah	Kilovolt-ampere-hour
kvar	Kilovar; reactive kilovolt-ampere
kvar hr	Kilovar-hour
kw	Kilowatt
KW	Kill(ed) well
kwh	Kilowatt hour
kwhm	Kilowatt-hourmeter

L

l	Liter
L/	Lower, as L/Gallup
/L	Line, as in E/L (East line)
LA	Level alarm, lightening avvester, load acid
Lab	Labor, laboratory
LACT	Lease automatic custody transfer
lad	Ladder
Lak	Lakota
L/Alb	Lower Albany
lam	Laminated, lamination(s)
La Mte	La Motte
Land	Landulina
Lans	Lansing
Lar	Laramie
LAS	Lower anhydrite stringer
lat	Latitude
Laud	Lauders
Layt	Layton
lb	Pound
LB	Light barrel
lb/ft	Pound per foot
lb-in	Pound-inch
LBOS	Light brown oil stain
lbr	Lumber
lb/sq ft	Pounds per square foot

LC	Lost circulation, long coupling, level controller, lease crude
LC	Lug cover type (5-gallon can)
lchd	Leached
LCL	Less-than-carload lot
LCM	Lost circulation material
L/Cret	Lower Cretaceous
LCP	Lug cover with pour spout
LCV	Level control valve
LD	Laid down
ld(s)	Land(s), load
LDC	Laid down cost
LDDCs	Laid (laying) down drill collars
LDDP	Laid (laying) down drill pipe
Leadv	Leadville
Le C	Le Comptom
LEL	Lower explosive limit
Len	Lennep
len	Lenticular
LFO	Light fuel oil
lg	Large, length, long, level glass
LGD	Lower Glen Dean
Lg Disc	Large Discorbis
Lge	League
LH	Left hand
LH/RP	Long handle/round point
LI	Level indicator
li	Lime, limestone
LIB	Light iron barrel
lic	License
LIC	Level indicator controller
Lieb	Liebuscella
lig	Lignite, lignitic
LIGB	Light iron grease barrel
LIH	Left in hole
lim	Limit, limonite
lin	Linear, liner
lin ft	Linear foot
liq	Liquid
liqftn	Liquefaction

litho	Lithographic
LJ	Lap joint
lk	Leak, lock
LKR	Locker
LLC	Liquid level controller
LLG	Liquid level gauge
lm	Lime
LMn	Lower Menard
Lmpy	Lumpy
LMTD	Log mean temperature difference
lmy	Limy
Lmy sh	Limy shale
LNG	Liquified natural gas
lngl	Linguloid
lnr	Liner
lns	Lense
LO	Load oil, lube oil
loc	Located, location
loc abnd	Location abandoned
loc gr	Location graded
long	Longitude(inal)
Lov	Lovington, Lovell
low	Lower
lp	Loop
LP	Low pressure, Lodge pole
L.P.	Line pipe
LP-Gas	Liquefied petroleum gas
LPO	Local purchase order
LP sep	Low-pressure separator
LR	Level recorder, long radius
LRC	Level recorder controller
lrg	Large
ls	Limestone
LSD	Legal subdivision (Canada), light steel drum
lse	Lease
lstr	Lustre
lt	Light
LT&C	Long threads & coupling
LTD	Log total depth

ltd	Limited
ltg	Lighting
LTL	Less than truck load
ltl	Little
ltr	Letter
LTS unit	Low-temperature separation unit
LTSD	Low-temperature shut-down
LTX unit	Low-temperature extraction unit
L/Tus	Lower Tuscaloosa
L U	Lease use (gas)
lub	Lubricant, lubricate (d) (ing) (ion)
lued	Lueders
LV	Liquid volume
LVI	Low viscosity index
lvl	Level
Lvnwth	Leavenworth
lwr	Lower
LW	Load water, lapweld

M

M/	Middle
m	Meter
MA	Massive Anhydrite, mud acid
ma	Milliampere
MAC	Medium amber cut
mach	Machine
Mack	Mackhank
Mad	Madison
mag	Magnetic, magnetometer
maint	Maintenance
maj	Major, majority
mall	Malleable
man	Manual, manifold
Manit	Manitoban
Mann	Manning
man op	Manually operated

Maq	Maquoketa
mar	Maroon, marine
March	Marchand
marg	Marginal
Marg	Marginulina
margas	Marine gasoline
Marg coco	Marginulina coco
Marg fl	Marginulina flat
Marg rd	Marginulina round
Marg tex	Marginulina texana
Mark	Markham
Marm	Marmaton
mass	Massive
Mass pr	Massilina pratti
mat	Matter
math	Mathematics
matl	Material
MAW	Mud acid wash
max	Maximum
May	Maywood
MB	Moody's Branch
MB	Methylene blue
Mbl Fls	Marble Falls
mbr	Member (geologic)
MBTU	Thousand British thermal units
MC	Mud cut
mc	Megacycle
MCA	Mud cleanout agent, mud-cut acid
McC	McClosky lime
McCul	McCullough
MCF	Thousand cubic feet
McEl	McElroy
MCFD	Thousand cubic feet per day
mchsm	Mechanism
McK	McKee
McL	McLish
MC Ls	Moore County Lime
McMill	McMillan
MCO	Mud-cut oil
mcr-x	Micro-crystalline

MCSW	Mud-cut salt water
MCW	Mud-cut water
MD	Measured depth
md	Millidarcies
MDDO	Maximum daily delivery obligation
MDF	Market demand factor
mdl	Middle
mdse	Merchandise
md wt	Mud weight
Mdy	Muddy
Meak	Meakin
meas	Measure (ed) (ment)
mech	Mechanic (al), mechanism
Mech DT	Mechanical down time
med	Median, medium
Med	Medina
Med B	Medicine Bow
med FO	Medium fuel oil
med gr	Medium-grained
Medr	Medrano
Meet	Meeteetse
MEG	Methane-rich gas
MEK	Methyl ethyl keton
memo	Memorandum
Men	Menard lime
Mene	Menefee
MEP	Mean effective pressure
MER	Maximum efficient rate
Mer	Meramec
merc	Mercury
mercap	Mercaptan
merid	Meridian
Meso	Mesozoic
meta	Metamorphic
meth	Methane
meth-bl	Methylene blue
meth-cl	Methyl chloride
methol	Methanol
metr	Metric
mev	Million electron volts

mezz	Mezzanine
MF	Manifold, mud filtrate
M&F	Male and female (joint)
MFA	Male to female angle
mfd	Microfarad, manufactured
mfg	Manufacturing
MFP	Maximum flowing pressure
M&FP	Maximum & final pressure
MG	Multi-grade, motor generator
mg	Milligram
m-gr	Medium-grained
m'gmt	Management
mgr	Manager
MH	Manhole
mh	Millihenry
MHz	Megahertz (megacycles per second)
MI	Malleable iron, mile(s), mineral interest, moving in (equipment)
mic	Mica, micaceous
micfos	Microfossil(iferous)
micro-xin	Microcrystalline
MICT	Moving in cable tools
MICU	Moving in completion unit
Mid	Midway
mid	Middle
MIDDU	Moving (moved) in double drum unit
MIK	Methyl isobutyl ketone
mil	Military, million
mill	Milliolitic
millg	Milling
MIM	Moving in materials
min	Minimum, minute(s), minerals
Minl	Minnelusa
min P	Minimum pressure
Mio	Miocene
MIPU	Moving in pulling unit
MIR	Moving in rig
MIRT	Moving in rotary tools
MIRU	Moving in and rigging up
misc	Miscellaneous

Mise	Misener
MISR	Moving in service rig
Miss	Mississippian
Miss Cany	Mission Canyon
MIST	Moving in standard tools
MIT	Moving in tools
MIU	Moisture, impurities and unsaponi-fiables (grease testing)
mix	Mixer
mkt	Market(ing)
Mkta	Minnekahta
mky	Milky
ML	Mud logger
ml	Milliliter
m/l	More or less
mld	Milled
mlg	Milling
ml TEL	Milliliters tetraethyl lead per gallon
MLU	Mud logging unit
MLW-PLAT	Mean low water to platform
mly	Marly
MM	Motor medium
mm	Millimeter
MMBTU	Million British thermal units
MMCF	Million cubic feet
MMCFD	Million cubic feet/day
mm Hg	Millimeters of mercury
MMSCFD	Million standard cubic feet per day
mnrl	Mineral
MO	Moving out, motor oil
mob	Mobile
MOCT	Moving out cable tools
MOCU	Moving out completion unit
mod	Moderate(ly), model, modification
modu	Modular
MOE	Milled other end
Moen	Moenkopi
mol	Molas, mollusca, mole
mol wt	Molecular weight
mon	Monitor

MON	Motor octane number
Mont	Montoya
Moor	Mooringsport
MOP	Maximum operating pressure
MOR	Moving out rig
Mor	Morrow
Morr	Morrison
MORT	Moving out rotary tools
Mos	Mosby
mot	Motor
mott	Mottled
MOU	Motor oil units
mov	Moving
Mow	Mowry
MP	Maximum pressure, melting point, multipurpose
MPB	Metal petal basket
MPGR-Lith	Multipurpose grease, lithium base
MPGR-soap	Multipurpose grease, soap base
MPH	Miles per hour
MPT	Male pipe thread
MR	Marine rig, meter run
mrlst	Marlstone
M & R Sta.	Measuring and regulating station
MS	Motor severe
MSA	Multiple service acid
MSP	Maximum surface pressure
mstr	Master
MT	Empty container, macaroni tubing
M/T	Marine terminal
MTD	Measured total depth, mean temperature difference
mtd	Mounted
mtg	Mounting
mtge	Mortgage
mtl	Material
MTP	Maximum top pressure, maximum tubing pressure
mtr	Meter
MTS	Mud to Surface

Mt. Selm	Mount Selman
M. Tus	Marine Tuscaloosa
mtx	Matrix
mudst	Mudstone
mud wt	Mud Weight
musc	Muscovite
mv	Millivolt
M/V	Motor vehicle, motor vessel
Mvde	Mesaverde
MVFT	Motor vehicle fuel tax
MW	Muddy water, microwave
MWD	Marine wholesale distributors
MWP	Maximum working pressure
MWPE	Mill wrapped plain end
Mwy	Midway
mxd	Mixed

N

N/2	North half
N/4	North quarter
NA	Not applicable, not available
Nac	Nacatoch
nac	Nacreous
NAG	No appreciable gas
NALRD	Northern Alberta land registration district
nap	Naphtha
nat	Natural
Nat'l	National
Nav	Navajo
Navr	Navarro
NB	Nitrogen blanket, new bit
Nbg	Newburg
NC	No core, National coarse thread, no change, normally closed
N. Cock	Nonionella Cockfieldensis

NCT	Non-contiguous tract
ND	Non-detergent, not drilling
NDBOPs	Nipple (d) (ing) down blowout preventers
Ndl Cr	Noodle Creek
NDT	Non-destructive testing
NE	Northeast, nonemulsifying agent
NE/4	Northeast quarter
NEA	Non-emulsion acid
NEC	Northeast corner
NEC	National Electric Code
neg	Negative, negligible
NEL	Northeast line
NEP	Net effective pay
neut	Neutral, neutralization
Neut. No.	Neutralization Number
New Alb	New Albany shale
Newc	Newcastle
NF	National Fine (thread), natural flow, no fluorescence, no fluid, no fuel
NFD	New field discovery
NFW	New field wildcat
NG	Natural gas, no gauge
NGL	Natural gas liquids
NGTS	No gas to surface
NIC	Not in contract
Nig	Niagara
Nine	Ninnescah
Niob	Niobrara
nip	Nipple
nitro	Nitrogylcerine
NL	North line
NL Gas	Non-leaded gas
N'ly	Northerly
NMI	Nautical mile
NO	New oil, normally open, number
NO	Noble-Olson
N/O	North offset
nod	Nodule, nodular
Nod Blan	Nodosaria blanpiedi

Nod mex	Nodosaria Mexicana
No Inc	No increase
NOJV	Non-operated joint ventures
nom	Nominal
Non	Nonionella
nonf G	Nonflammable compressed gas
NOP	Non-operating property
nor	Normal
NOR	No order required
no rec	No recovery
noz	Nozzle
NP	Nameplate, notary public, nickle plated, not prorated, no production, not pumping, non-porous
NPD	New pool discovery
npne	Neoprene
NPOS	No paint on seams
NPS	Nominal pipe size
NPT	National pipe thread
NPTF	National pipe thread, female
NPTM	National pipe thread, male
NPW	New pool wildcat
NPX	New pool exempt (nonprorated)
NR	No report, not reported, no recovery, non-returnable, no returns
NRS	Non-rising stem (valve)
NRSB	Non-returnable steel barrel
NRSD	Non-returnable steel drum
NS	No show
NSG	No show gas
NSO	No show oil
NSO&G	No show oil & gas
nstd	Non-standard
N/S S/S	Non-standard service station
NT	Net tons, no time
NTD	New total depth
NTS	Not to Scale
N/tst	No test
NU	Non-upset, nippling up
NUBOPs	Nipple (d) (ing) up blowout preventers

NUE	Non-upset ends
Nug	Nugget
num	Numerous
NVP	No visible porosity
NW	Northwest, no water
NW/C	Northwest corner
NW/4	Northwest quarter
NWL	Northwest line
NWT	Northwest Territories
NYA	Not yet available

O

O	Oil
OA	Overall
OAH	Overall height
Oakv	Oakville
OAL	Overall length
OAW	Old abandoned well
OB	Off bottom
obj	Object
OBM	Oil base mud
OBMO	Outboard motor oil
OBS	Ocean bottom suspension
obsol	Obsolete
OBW & RS	Optimum bit weight and rotary speed
OC	Oil cut, on center, open cup, operations commenced
O/C	Oil change
OCB	Oil circuit breaker
occ	Occasional(ly)
OCM	Oil-cut mud
OCS	Outer continental shelf
OCSW	Oil-cut salt water
oct	octagon, octagonal, octane
OCW	Oil-cut water

OD	Outside diameter
od	Odor
Odel	O'Dell
OE	Oil emulsion, open end
OEB	Other end beveled
OEM	Oil emulsion mud
OF	Open flow
off	Office, official
off-sh	Off-shore
OFL	Overflush(ed)
OFLU	Oil fluorescence
OFOE	Orifice flange one end
OFP	Open flow potential
O&G	Oil and gas
O&GCM	Oil and gas-cut mud
O&GC SULW	Oil & gas-cut sulphur water
O&GCSW	Oil and gas-cut salt water
O&GCW	Oil and gas-cut water
OGJ	Oil and Gas Journal
O&GL	Oil and gas lease
OH	Open hearth, open hole, over head
O'H	O'Hara
ohm cm	Ohm-centimeter
ohm-m	Ohm-meter
OIH	Oil in hole
Oil Cr	Oil Creek
OIP	Oil in place
ole	Olefin
Olig	Oligocene
ONR	Octane number requirement
ONRI	Octane number requirement increase
OO	Oil odor
ooc	Oolicastic
ool	Oolitic
oom	Oolimoldic
OP	Oil pay, over produced, out post
OPBD	Old plug-back depth
oper	Operate, operations, operator
Operc	Operculinoides

opn	Open (ing) (ed)
opp	Opposite
OPI	Oil payment interest
OPT	Official potential test
optn to F/O	Option to farmout
Or	Oread
Ord	Ordovician
orf	Orifice
org	Organic, organization
orig	Original, originally
Orisk	Oriskany
ORR	Overriding royalty
ORRI	Overriding royalty interest
orth	Orthoclase
Os	Osage
OS	Oil show, overshot
O/S	Out of service, over and short (report), out of stock
OSA	Oil soluble acid
Osb	Osborne
O sd	Oil sand
OSF	Oil string flange
O S & F	Odor, stain & fluorescence
OSI	Oil well shut in
Ost	Ostracod
OSTN	Oil stain
OSTOIP	Original stock tank oil in place
Osw	Oswego
O&SW	Oil and salt water
O&SWCM	Oil & sulphur water-cut mud
OS&Y	Outside screw and yoke (valve)
OT	Open tubing, overtime
OTD	Old total depth
OTE	Oil-powered total energy
otl	Outlet
OTS	Oil to surface
O T & S	Odor, taste & stain
O, T, S & F	Odor, taste, stain & fluorescence
OU	Oil unit
Our	Ouray

ovhd	Overhead
O & W	Oil and water
OWC	Oil-water contact
OWDD	Old well drilled deeper
OWF	Oil well flowing
OWG	Oil-well gas
OWPB	Old well plugged back
OWWO	Old well worked over
ox	Oxidized, oxidation
oxy	Oxygen
oz	Ounce

P

P & A	Plugged & abandoned
PA	Pooling agreement, pressure alarm
PAB	Per acre bonus
Padd	Paddock
Paha	Pahasapa
Pal	Paluxy
Paleo	Paleozoic, paleontology
Palo P	Palo Pinto
Pan L	Panhandle Lime
PAR	Per acre rental
Para	Paradox
Park C	Park City
pat	Patent(ed)
patn	Pattern
pav	Paving
Paw	Pawhuska
payt	Payment
PB	Plugged Back
PBD	Plugged back depth
PBHL	Proposed bottom hole location
pbl	Pebble
pbly	Pebbly
PBP	Pulled big pipe

PBTD	Plugged back total depth
PBW	Pipe, buttweld
PBX	Private branch exchange
PC	Paint Creek, poker chipped, Porter Creek
P&C	Personal and confidential
pc	Piece
pcs	Pieces
pct	Percent
PCV	Pressure control valve, positive crankcase ventilation
PD	Per day, proposed depth, pressed distillate, paid
PDC	Pressure differential controller
PDET	Production department exploratory test
PDI	Pressure differential indicator
PDIC	Pressure differential indicator controller
PDR	Pressure differential recorder
PDRC	Pressure differential recorder controller
pdso	Pseudo
PE	Plain end, pumping equipment
PEB	Plain end beveled
pebs	Pebbles
pell	Pelletal, pelletoidal
pen	Penetration, penetration test
Pen A.C.	Penetration asphalt cement
penal	Penalize, penalized, penalizing, penalty
Penn	Pennsylvanian
perco	Percolation
perf	Perforate (d) (ing) (or)
perf csg	Perforated casing
perm	Permeable, permeability, permanent
Perm	Permian
perp	Perpendicular
pers	Personnel
pet	Petroleum

Pet	Pettet
petrf	Petroliferous
petrochem	Petrochemical
Pet sd	Pettus sd
Pett	Pettit
PEW	Pipe, electric weld
PF	Power factor
P&F	Pump and flow
pfd	Preferred
PFM	Power factor meter
PFT	Pumping for test
PG	Pecan Gap
PGC	Pecan Gap Chalk
PGW	Producing gas well
ph	Phase
Ph	Parish
Phos	Phosphoria
PI	Penetration index, Pine Island, pressure indicator, productivity index
PIC	Pressure indicator controller
Pic Cl	Pictured Cliff
pinpt	Pinpoint
PIP	Pump-in pressure
piso	Pisolites, pisolitic
pit	Pitted
PJ	Pump jack, pump job
pk	Pink
pkd	Packed
pkg	Packing, package
pkgd	Packaged
pkr	Packer
PL	Pipeline, property line
P&L	Profit & loss
plag	Plagioclase
Plan. harangensis	Planulina harangensis
Plan. palm.	Planulina palmarie
P Lar	Post Laramie
plas	Plastic
platf	Platform

platfr	Platformer
plcy	Pelecypod
pld	Pulled
PLE	Plain large end
Pleist	Pleistocene
pl fos	Plant fossils
plg	Pulling
plgd	Plugged
Plio	Pliocene
pln	Plan
plngr	Plunger
PLO	Pipeline oil, pumping load oil
plt	Plant
PLT	Pipeline terminal
plt	Pilot
plty	Platy
PLW	Pipe, lapweld
PM	Pensky Martins
pmp (d) (g)	Pump, pumped, pumping
PN	Performance Number (Av gas)
pneu	Pneumatic
P & NG	Petroleum & natural gas
pnl	Panel
PNR	Please note and return
P. O.	Post Oak, Pin Oak
PO	Pumps off, purchase order, pulled out
po	Phrrhotite
POB	Plug on bottom, pump on beam
Pod.	Podbielniak
POE	Plain one end
POGW	Producing oil & gas well
POH	Pulled out of hole
pois	Poison
pol	Polish(ed)
poly	Polymerization, polymerized
poly cl	Polyvinyl chloride
polyel	Polyethylene
polygas	Polymerized gasoline
polypl	Polypropylene

PONA	Paraffins, olefins, naphthenes, aromatics
Pont	Pontotoc
POP	Putting on pump
por	Porosity, porous
porc	Porcelaneous, Porcion
port	Portable
pos	Position, positive
poss	Possible(ly)
pot	Potential
pot dif	Potential difference
pour ASTM	Pour point (ASTM Method)
POW	Producing oil well
POWF	Producing oil well flowing
POWP	Producing oil well pumping
PP	Pinpoint, production payment, pulled pipe
P&P	Porosity and permeability, porous and permeable
ppd	Prepaid
ppg	Pounds per gallon
PPI	Production payment interest
ppm	Parts per million
PPP	Pinpoint porosity
ppt	Precipitate
pptn No	Precipitation Number
PR	Polished rod, public relations, pressure recorder
pr	Pair
PRC	Pressure recorder control
prcst	Precast
prd	Period
Pre Camb	Pre-Cambrian
predom	Predominant
prefab	Prefabricated
prehtr	Preheater
prelim	Preliminary
prem	Premium
Prep	Prepare, preparing, preparation
press	Pressure

prest	Prestressed
prev	Prevent, preventive
PRF	Primary Reference Fuel
pri	Primary
prin	Principal
pris	Prism(atic)
priv	Privilege
prly	Pearly
prmt	Permit
prncpl lss	Principal lessee(s)
pro	prorated
prob	Probable(ly)
proc	Process
prod	Produce (d) (ing) (tion), product(s)
prog	Progress
proj	Project (ed) (ion)
prop	Proportional, proposed(d)
prot	Protection
Protero	Proterozoic
Prov	Provincial
PRPT	Preparing to take potential test
PR&T	Pull(ed) rods & tubing
prtgs	Partings
partly	Partly
PS	Pressure switch
ps	Pseudo
PSA	Packer set at
PSD	Permanently shut down
PSE	Plain small end
psf	Pounds per square foot
psi	Pounds per square inch
PSI	Profit sharing interest
psia	Pounds per square inch absolute
psig	Pounds per square inch gauge
PSL	Public School Land
PSM	Pipe, seamless
PSW	Pipe, spiral weld
PT	Potential test
pt	Part, partly, point, pint
PTG	Pulling tubing

Pk Lkt	Point Lookout
PTR	Pulling tubing and rods
PTS pot.	Pipe to soil potential
PTTF	Potential test to follow
PU	Picked up, pulled up, pumping unit
purp	Purple
PV	Plastic viscosity, pore volume
PVC	Polyvinyl chloride
pvmnt	Pavement
PVR	Plant volume reduction
PVT	Pressure-volume-temperature
PWR	Power
Pxy	Paluxy
pyls	Pyrolysis
pyr	Pyrite, pyritic
pyrbit	Pyrobitumen
pyrclas	Pryoclastic

Q

Q. City	Queen City
QDA	Quantity discount allowance
qnch	Quench
QRC	Quick ram change
qry	Quarry
Q. Sd	Queen Sand
qt	Quart(s)
qtr	Quarter
qtz	Quartz, quartzite, quartzitic
qtzose	Quartzose
qty	Quantity
quad	Quadrant, quadrangle, quadruple
qual	Quality
quan	Quantity
quest	Questionable
quint	Quintuplicate

R

R	Range, Rankine (temp. scale)
RA	Right angel, radioactive
R/A	Regular acid
rad	Radial, radian, radiological, radius
radtn	Radiation
RALOG	Running radioactive log
Rang	Ranger
RB	Rotary bushing, rock bit
Rbls	Rubber balls
RBM	Rotary bushing measurement
RBP	Retrievable bridge plug
rbr	Rubber
RBSO	Rainbow show of oil
RBSOF	Rubber ball sand oil frac
RBSWF	Rubber ball sand water frac
RC	Rapid curing, remote control, reverse circulation, running casing, Red Cave
RCO	Returning circulation oil
RCR	Ramsbottom Carbon Residue, reverse circulation rig
RD	Rigged down, rigging down
rd	Round, road
RDB	Rotary drive bushing
Rd Bds	Red Beds
RDB-GD	Rotary drive bushing to ground
rdd	Rounded
Rd Fk	Red Fork
Rd Pk	Red Peak
rds	Roads
RDSU	Rigged down swabbing unit
rd thd	Round thread
rdtp	Round trip
R & D	Research and development
reac	Reactor
reacd	Reacidize, reacidizing, reacidized

Reag	Reagan
rebar	Reinforcing bar
reblr	Reboiler
rec	Recover, recovering, recovered, recovery, recommend, recorder, recording
recd	Received
recip	Reciprocate(ing)
recirc	Recirculate
recomp	Recomplete (d) (ion)
recond	Recondition(ed)
recp	Receptacle
rect	Rectifier, rectangle, rectangular
recy	Recycle
red	Reducing, reducer
red bal	Reducing balance
redrld	Redrilled
ref	reference, refine (d) (r) (ry)
refer	Refrigeration
referg	Refrigerant
refg	Refining
refgr	Refrigerator
refl	Reflection
refl	Reflux
reform	Reformate, reformer, reforming
refr	Refraction, refractory
reg	Regular, regulator, register
regen	Regenerator
reinf	Reinforce (d) (ing)
reinf conc	Reinforced concrete
rej	Reject
rej'n	Rejection
Rek	Reklaw
rel	Relay, release(d)
REL	Running electric log
reloc	relocate(d)
rem	Remains, remedial, remove (al) (able)
Ren	Renault
rent	Rental
Reo bath	Reophax bathysiphoni

rep	Repair (ed) (ing) (s), replace (d), report
reperf	Reperforated
repl	Replace(ment)
req	Requisition
reqd	Required
reqmt	Requirement
reqn	Requisition
res	Research, reservation, reserve, reservoir, resistance, resistivity, resistor
resid	Residual, residue
Res. O. N.	Research Octane Number
ret	Retain (ed) (er) (ing), retard(ed), return
retd	Returned
retr ret	Retrievable retainer
rev	Reverse(d), revise (d) (ing) (ion), revolution(s)
rev/O	Reversed out
RF	Raised faced, rig floor
RFFE	Raised face flanged end
RFWN	Raised face weld neck
RG	Ring groove
rg	Ring
Rge	Range
rgh	Rough
RH	Rat hole, right hand
RHD	Right hand door
rheo	Rheostat
RHM	Rat hole mud
RHN	Rockwell hardness number
RI	Royalty interest
Rib	Ribbon sand
Rier	Rierdon
rig rel	Rig released
RIH	Ran in hole
RIL	Red indicating lamp
riv	Rivet
RJ	Ring Joint

RJFE	Ring joint flanged end
rk	Rock
rky	Rocky
RL	Random lengths
R&L	Road & location
R&LC	Road & location complete
rlf	Relief
rlg	Railing
rls (d) (ing)	Release (d) (ing)
rly	Relay
rm	Room, ream
rmd	Reamed
rmg	Reaming
rmn	Remains
RMS	Root mean square
rmv	Removable
rnd	Rounded
rng	Running
RO	Reversed out
Ro	Rosiclare sand
R.O.	Red Oak
R&O	Rust & oxidation
ro	Rose
Rob	Robulus
Rod	Rodessa
ROF	Rich oil fractionator
ROI	Return on investment
ROL	Rig on location
ROM	Run of mine
RON	Research Octane Number
ROP	Rate of penetration
ROR	Rate of return
rot	Rotary, rotate, rotator
ROW	Right-of-way
roy	Royalty
RP	Rock pressure
rpm	Revolutions per minute
rpmn	Repairman
RPP	Retail pump price
rps	Revolutions per second

RR	Railroad, Red River, rig released
RRC	Railroad Commission (Texas)
RR&T	Ran (running) rods and tubing
RS	Rig skidded, rising stem (valve)
RSD	Returnable steel drum
RSH	Mercaptan
rsns	Resinous
RSU	Released swab unit
RT	Rotary tools, rotary table
R & T	Rods & tubing
R test	Rotary test
RTG	Running tubing
rtg	Rating
rthy	Earthy
RTJ	Ring tool joint, ring type joint
RTLTM	Rate too low to measure
rtnr	Retainer
RTTS	Retrievable test treat squeeze (tool)
RU	Rigging up, rigged up
RU	Rotary unit
rub	Rubber
RUCT	Rigging up cable tools
RUM	Rigging up machine
RUP	Rigging up pump
rupt	Rupture
RUR	Rigging up rotary
RURT	Rigging up rotary tools
RUSR	Rigging Up service rig
RUST	Rigging up standard tools
RUT	Rigging up tools
RVP	Reid vapor pressure
rvs(d)	Reverse(d)
R/W	Right of way
rwk(d)	Rework(ed)
RWTP	Returned well to production
Ry	Railway

S

S/2	South half
S/	Swabbed
Sab	Sabinetown
sach	Saccharoidal
Sad Cr	Saddle Creek
sadl	Saddle
saf	Safety
sal	Salary, salaried, salinity
Sal	Salado
Sal Bay	Saline Bayou
salv	Salvage
samp	Sample
Sana	Sanastee
San And	San Andres
San Ang	San Angelo
sani	Sanitary
San Raf	San Rafael
sap	Saponification
Sap No.	Saponification number
Sara	Saratoga
sat	Saturated, saturation
Saw	Sawatch
Sawth	Sawtooth
Say Furol	Saybolt Furol
SB	Sideboom, sleeve bearing, stuffing box
Sb	Sunburst
sb	Sub
SBA	Secondary butyl alcohol
SBB&M	San Bernardino Base and Meridian
SBHP	Static bottom-hole pressure
S Bomb	Sulfur by bomb method
SC	Show condensate
sc	Scales
SCF	Standard cubic foot
SCFD	Standard cubic feet per day

SCFH	Standard cubic feet per hour
SCFM	Standard cubic feet per minute
sch	Schedule
schem	Schematic
scly	Securaloy
scolc	Scolescodonts
scr	Scratcher, screw, screen
scrd	Screwed
scrub	Scrubber
sctrd	Scattered
SD	Shut down
sd	Sand, sandstone
SDA	Shut down to acidize
SD Ck	Side door choke
SDF	Shut down to fracture
sdfract	Sandfract
SDL	Shut down to log
SDO	Show of dead oil
SDO	Shut down for orders
sdoilfract	Sand oil fract
SDON	Shut down overnight
SDPA	Shut down to plug & abandon
SDPL	Shut down for pipe line
SDR	Shut down for repairs
Sd SG	Sand showing gas
sd & sh	Sand and shale
Sd SO	Sand showing oil
sdtkr	Sidetrack (ed) (ing)
SDW	Shut down for weather
sdwtrfract	Sand water fract
SDWO	Shut down awaiting orders
sdy	Sandy
sdy li	Sandy lime
sdy sh	Sandy shale
SE	Southeast
SE/4	Southeast quarter
S/E	Screwed end
Sea	Seabreeze
sec	Secant, second, secondary, secretary, section

SE/C	Southeast corner
sed	Sediment(s)
Sedw	Sedwick
seis	Seismograph, seismic
Sel	Selma
sel	Selenite
Sen	Senora
SE NA	Screw end American National Acme thread
SE NC	Screw end American National Coarse thread
SE NF	Screw end American National Fine thread
SE No.	Steam Emulsion Number
SE NPT	Screw End American National Taper Pipe Thread
sep	Separator
sept	Septuplicate
seq	Sequence
ser	Series, serial
Serp	Serpentine
Serr	Serratt
serv	Service(s)
serv chg	Service change
set	Settling
sew	Sewer
Sex	Sexton
sext	Sextuplicate
SF	Sandfrac
S&F	Swab and flow
sfc	Surface
SFL	Starting fluid level
SFLU	Slight, weak, or poor fluorescence
SFO	Show of free oil
sft	Soft
SG	Show gas, surface geology
SG&C	Show gas and condensate
SGCM	Slightly gas-cut mud
SGCO	Slightly gas-cut oil
SGCW	Slightly gas-cut water

SGCWB	Slightly gas-cut water blanket
SG&D	Show gas and distillate
sgd	Signed
sgls	Singles
SG&O	Show of gas and oil
SG&W	Show gas & water
sh	Shale, sheet
Shan	Shannon
SHDP	Slim hole drill pipe
Shin	Shinarump
shls	Shells
shld	Shoulder
shly	Shaley
shp	Shaft horsepower
shpg	Shipping
shpt	Shipment
shr	Shear
SHT	Straight hole test
shthg	Sheathing
SI	Shut in
SIBHP	Shut in bottom hole pressure
SICP	Shut in casing pressure
sid	Siderite(ic)
SIGW	Shut in gas well
Sil	Silurian
silic	Silica, siliceous
silt	Siltstone
sim	Similar
Simp	Simpson
SIOW	Shut in oil well
SIP	Shut in pressure
siph. d.	Siphonina davisi
SITP	Shut in tubing pressure
SIWHP	Shut in well head pressure
SIWOP	Shut in-waiting on potential
sk	Sacks
Sk Crk	Skull Creek
skim	Skimmer
Skn	Skinner
sks	Slickensided

skt	Socket
SL	Section line, state lease, south line
sl	Sleeve, slight(ly)
SLC	Steel line correction
sld	Sealed
Sli	Sligo
sli	Slight(ly)
sli SO	Slight show of oil
slky	Silky
SLM	Steel line measurement
slnd	Solenoid
slt	Silt
Slt Mt	Salt Mountain
Slty	Salty
slur	Slurry
SM	Surface measurement
sm	Small
Smithw	Smithwick
Smk	Smackover
smls	Seamless
smth	Smooth
SN	Seating nipple
S O	South offset, shake out, show oil, slip on, side opening
S&O	Stain and odor
SOCM	Slightly oil-cut mud
SOCW	Slight oil-cut water
sod gr	Sodium base grease
SOE	Screwed on one end
SOF	Sand oil fracture
SO&G	Show oil and gas
SO&GCM	Slightly oil & gas-cut mud
SOH	Shot open hole
sol	Solenoid, solids
soln	Solution
solv	Solvent
som	Somastic
somct	Somastic coated
SOP	Standard operational procedure
sort	Sorted(ing)

SO&W	Show oil and water
sow	Socket weld
SP	Self (Spontaneous) potential, set plug, surface pressure, straddle packer, shot point, slightly porous
Sp	Sparta
sp	Spare, spore
s&p	Salt & pepper
spcl	Special
spcr	Spacer
spd	Spud (ded) (der)
spdl	Spindle
SPDT	Single pole double throw
SP–DST	Straddle packer drill stem test
spec	Specification
speck	Speckled
spf	Spearfish
sprf	Spirifers
spg	Sponge, spring
sp gr	Specific gravity
sph	Spherules
Sphaer	Sphaerodina
sphal	Sphalerite
sp ht	Specific heat
spic	Spicule(ar)
Spiro, b.	Spiroplectammina barrowi
spkt	Sprocket
spkr	Sprinkler
splty	Splintery, split
Spletp	Spindletop
splty	Specialty
sply	Supply
SPM	Strokes per minute
Spra	Spraberry
Sprin	Springer
S Riv	Seven Rivers
SPST	single pole single throw
SPT	Shallower pool (pay) test
sptd	Spotted

sptty	Spotty
sp. vol.	Specific volume
sq	Square, squeezed
sq cg	Squirrel cage
sq cm	Square centimeter
sq ft	Square foot
sq in	Square inch
sq km	Square kilometer
sq m	Square meter
sq mm	Square millimeter
sq pkr	Squeeze packer
sq yd	Square yard(s)
sqz	Squeeze (d) (ing)
SR	Short radius
SRL	Single random lengths
srt (d) (g)	Sort (ed) (ing)
SS	Stainless steel, service station, single shot, slow set (cement), string shot, subsea, subsurface, small show
s & s	Spigot and spigot
SSG	Slight show of gas
SSO	Slight show oil
SSO&G	Slight show of oil & gas
S/SR	Sliding scale royalty
SSU	Saybolt Seconds Universal
SSUW	Salty sulfur water
ST	Short thread
ST(g)	Sidetrack(ing)
S/T	Sample tops
sta	Station
stab	Stabilized(er)
Stal	Stalnaker
Stan	Stanley
stat	Stationary, statistical
State pot	State potential
STB	Stock tank barrels
STB/D	Stock tank barrels per day
s,t&b	Sides, tops & bottoms
ST&C	Short threads & Coupling

stcky	Sticky
std (s) (g)	Standard, stand (s) (ing)
stdy	Steady
Stel	Steele
steno	Stenographer
Stens	Stensvad
St. Gen	Saint Genevieve
stging	Straightening
STH	Sidetracked hole
stip	Stippled
stir	Stirrup
stk	Stock, stuck, streaks, streaked
St L	Saint Louis Lime
stl	Steel
STM	Steel tape measurement
stm	Steam
stm cyl oil	Steam cylinder oil
stm eng oil	Steam engine oil
stn (d) (g)	Stain (ed) (ing)
stn/by	Stand by
Stn Crl	Stone Corral
Stnka	Satanka
stnr	Strainer
stoip	Stock tank oil in place
stor	storage
STP	Standard temperature and pressure
stp	Stopper
stpd	Stopped
St Ptr	Saint Peter
S-T-R	Section-township-range
Str	Strawn
strat	Stratigraphic
strd	Straddle, strand(ed)
strg	Strong, storage, stringer
stri	Striated
strom	Stromatoporoid
strt	Straight
struc	Structure, structural
STTD	Sidetracked total depth
stv	Stove oil

stwy	Stairway
styo	Styolite, styolitic
Sty Mt	Stony Mountain
sub	Subsidiary, substance
Sub Clarks	Sub-Clarksville
subd	Subdivision
substa	Substation
suc	Sucrose, sucrosic
suct	Suction
sug	Sugary
sul	Sulphur (sulfur)
sulph	Sulphated
sul wtr	Sulphur water
sum	Summary
Sum	Summerville
Sunb	Sunburst
Sund	Sundance
Sup	Supai
supl	Supply (ied) (ier) (ing)
supp	supplement
suppt	Support
suprv	Supervisor
supsd	Superseded
supt	Superintendent
sur	Survey
surf	Surface
surp	Surplus
SUS	Saybolt Universal Seconds
susp	Suspended
svc	Service
svcu	Service unit
SVI	Smoke Volatility Index
Svry	Severy
SW/4	Southwest quarter
SW	salt wash, salt water, spiral weld, socket weld, southwest
Swas	Swastika
swbd	Switchboard
swb (d) (g)	Swabbed, swabbing
SWC	Sidewall cores

SW/c	Southwest corner
SWD	Salt water disposal
swd	Swaged
SWDS	Salt water disposal system
SWDW	Salt water disposal well
swet	Sweetening
SWF	Sand-water fracture
swg	Swage
swgr	Switchgear
SWI	Salt water injection
SWP	Steam working pressure
SWS	Sidewall samples
SWTS	Salt water to surface
SWU	Swabbing unit
sx	Sacks
sxtu	Sextuple
Syc	Sycamore
Syl	Sylvan
sym	Symbol, symmetrical
syn	Synthetic, synchronous, synchronizing
syn conv	Synchronous converter
sys	System
sz	Size

T

T	Tee
T	Ton (after a number)
T	Township (as T2N)
T/	Top of (a formation)
TA	Temporarily abandoned, turn around
tab	Tabular, tabulating
Tag	Tagliabue
Tal	Tallahatta
Tamp	Tampico

Tan	Tansill
Tann	Tannehill
Tark	Tarkio
Tay	Taylor
TB	Tank battery, thin bedded
T & B	Top and bottom
tb	Tube
TBA	Tertiary butyl alcohol; tires, batteries, and accessories
T&BC	Top & bottom chokes
TBE	Threaded both ends
tbg	Tubing
tbg chk	Tubing choke
tbg press	Tubing pressure
TBP	True boiling point
TC	Temperature controller, tool closed, top choke, tubing choke
T/C	Tank car
T & C	Threaded & coupled, topping and coking
TCC	Thermofor catalytic cracking
TCP	Tricresyl phosphate
TCV	Temperature control valve
TD	Total depth
TDA	Temporary dealer allowance
TDI	Temperature differential indicator
TDR	Temperature differential recorder
tech	Technical, technician
TEFC	Totally enclosed-fan cooled
tel	Telephone, telegraph
TEL	Tetraethyl lead
Tel Cr	Telegraph Creek
Temp	Temperature, temporary (ily)
Tens	Tensleep
Tent	Tentaculites
tent	Tentative
Ter	Tertiary
term	Terminal
termin	Terminate (d) (ing) (ion)
Tex	Texana

tex	Texture
Text. art.	Textularia articulate
Text. d.	Textularia dibollensis
Text. h.	Textularia hockleyensis
Text. w.	Textularia warreni
Tfing	Three Finger
Tfks	Three Forks
tfs	Tuffaceous
T&G	Tongue and groove (joint)
tgh	Tough
TH	Tight hole
th	Thence
Thay	Thaynes
thd	Thread, threaded
Ther	Thermopolis
therm	Thermometer
therst	Thermostat
THF	Tubinghead flange
THFP	Top hole flow pressure
thk	Thick, thickness
thrling	Throttling
thrm	Thermal
thrm ckr	Thermal cracker
thru	Through
Thur	Thurman
TI	Temperature indicator
ti	Tight
TIC	Temperature indicator controller
TIH	Trip in hole
Tim	Timpas
Timpo	Timpoweap
tk	Tank
tkg	Tankage
tkr	Tanker(s)
TLE	Thread large end
TLH	Top of liner hanger
tl	Tool, tools
TML	Tetramethyl lead
tndr	Tender
TNS	Tight no show

TO	Tool open
TOBE	Thread on both ends
TOC	Top of cement
TOCP	Top of cement plug
Tod	Todilto
TOE	Threaded one end
TOF	Top of fish
TOH	Trip out of hole
tol	Tolerance
TOL	Top of liner
tolu	Toluene
Tonk	Tonkawa
TOP	Testing on pump
topg	Topping
topo	Topographic, topography
TOPS	Turned over to producing section
Tor	Toronto
Toro	Toroweap
TORT	Tearing out rotary tools
tot	Total
Tow	Towanda
TP	Travis Peak, tubing pressure, tool pusher
T/pay	Top of pay
TPC	Tubing pressure–closed
TPF	Threaded pipe flange, tubing pressure-flowing
tpk	Turnpike
Tpka	Topeka
TPSI	Tubing pressure shut in
TR	Temperature recorder
tr	Tract, trace
T&R	Tubing and rods
trans	Transformer
trans	Transfer (ed) (ing), transmission
transl	Translucent
transp	Transparent, transportation
TRC	Temperature recorder controller
Tren	Trenton
Tremp	Teremplealeau

Tri	Triassic
trilo	Trilobite
Trin	Trinidad
trip	Tripoli, tripolitic, triplicate, tripped (ing)
trkg	Trackage
trk	Truck
Trn	Trenton
trt (d) (g)	Treat (ed) (ing)
trtr	Treater
TS	Tensile strength, Tar Springs sand
T/S	Top salt
TSD	Temporarily shut down
T/sd	Top of sand
TSE	Thread small end
TSE-WLE	Thread small end, weld large end
TSI	Temporarily shut in
TSITC	Temperature Survey indicated top cement at
tst (d) (g)	Test (ed) (ing)
tste	Taste
TSTM	Too small to measure
tstr	Tester
TT	Tank truck, through tubing
TTF	Test to follow
TTL	Total time lost
TTTT	Turned to test tank
Tuck	Tucker
tuf	Tuffaceous
Tul Cr	Tulip Creek
tung carb	Tungsten carbide
Tus	Tuscaloosa
TV	Television
TVA	Temporary voluntary allowance
TVD	True vertical depth
TVP	True Vapor Process
TW	Tank wagon
T&W	Tarred and wrapped
Tw Cr	Twin Creek
twp	Township

twst	Townsite
twst off	Twisted off
TWTM	Too weak to measure
TWX	Teletype
ty	Type
typ	Typical
tywr	Typewriter

U

U/	Upper (i.e., U/Simpson)
U/C	Under construction
UCH	Use customer's hose
UD	Under digging
UG	Under gauge, underground
UGL	Universal gear lubricant
UHF	Ultra high frequency
U/L	Upper and lower
ult	Ultimate
un	Unit
unbr	Unbranded
unconf	Unconformity
uncons	Unconsolidated
undiff	Undifferentiated
unf	Unfinished
uni	Uniform
univ	University, universal
UR	Under reaming, unsulfonated residue
UV	Union Valley
Uvig. lir.	Uvigerina lirettensis
U/W	Used with

V

V	Volume
v	Volt
v.	Very (as very tight)
va	Volt-ampere
vac	Vacuum, vacant, vacation
Vag. reg	Vaginuline regina
Val	Valera
Vang	Vanguard
vap	Vapor
var	Variable, various, volt-ampere reactive
vari	Variegated
v. c.	Very common
VCP	Vitrified clay pipe
vel	Velocity
vent	Ventilator
Ver Cl	Vermillion Cliff
Verd	Verdigris
vert	Vertical
ves	Vesicular
v-f-gr	Very fine-grained
VHF	Very high frequency
v-HOCM	Very heavily oil-cut mud
VI	Viscosity index
Vi	Viola
Virg	Virgelle
vis	Viscosity, visible
vit	Vitreous
Vks	Vicksburg
V/L	Vapor-liquid ratio
VLAC	Very light amber cut
vlv	Valve
VM&P Naphtha	Varnish makers & painters naphtha
v. n.	Very noticeable
Vogts	Vogtsberger
vol	Volume

vol. eff.	Volumetric efficiency
VP	Vapor pressure
V.P.S.	Very poor sample
v. r.	Very rare
vrs	Varas
vrtb	Vertebrate
vrtl	Vertical
vrvd	Varved
V/S	Velocity survey
vs	Versus
VSGCM	Very slight gas-cut mud
v-sli	Very slight
VSP	Very slightly porous
VSSG	Very slight show of gas
VSSO	Very slight show of oil
vug	Vuggy, vugular

W

w	Watt
W	West, wall (if used with pipe)
W/2	West half
w/	With
Wa Sd	Waltersburg sand
WAB	Weak air blow
Wab	Wabaunsee
Wad	Waddell
Wap	Wapanucka
War	Warsaw
Was	Wasatch
Wash	Washita
WB	Water blanket, wet bulb, Woodbine
WBIH	Went back in hole
WC	Wildcat, water cushion (DST), Wolfe City, water cut
WCM	Water-cut mud
WCO	Water-cut oil

W Cr	Wall Creek
WCTS	Water cushion to surface
WD	Water depth
WD	Water disposal well
Wdfd	Woodford
Wd R	Wind River
WE	Weld ends
Web	Weber
Well	Wellington
WF	Waterflood, wide flange
W-F	Washita-Fredericksburg
WFD	Wildcat field discovery
wgt.	Weight
WH	Wellhead
Wh Dol	White Dolomite
whip	Whipstock
Wh Sd	White sand
whse	Warehouse
whsle	Wholesale
wht	White
WI	Washing in, water injection, working interest, wrought iron
Wich.	Wichita
Wich Alb	Wichita Albany
WIH	Water in hole, went in hole
Willb	Willberne
Win	Winona
Winf	Winfield
Wing	Wingate
Winn	Winnipeg
wk	Weak, week
wkd	Worked
wkg	Working
wko	Workover
wkor	Workover rig
WL	West line, wire line, water loss
W/L	Water load
WLC	Wire line coring
wld	Welded, welding
wldr	Welder
WLT	Wireline test

WLTD	Wireline total depth
W'ly	Westerly
WN	Weld neck, welding neck
WNSO	Water not shut-off
WO	Waiting on
WO	Workover, wash over, work order
W/O	West offset, without
WOA	Waiting on allowable, waiting on acid
WOB	Waiting on battery
W.O.B.	Weight on bit
WOC	Waiting on cement
WOCR	Waiting on completion rig
WOCT	Waiting on cable tools, or completion tools
WODP	Without drill pipe
WOG	Water, oil or gas
Wolfc	Wolfcamp
WOO	Waiting on orders
Wood	Woodside
Woodf	Woodford
WOP	Waiting on permit, waiting on pipe, waiting on pump
WOPE	Waiting on production equipment
WOPT	Waiting on potential test
WOPU	Waiting on pumping unit
WOR	Waiting on rig or rotary, water-oil ratio
WORT	Waiting on rotary tools
WOS	Washover string
WOSP	Waiting on state potential
WOST	Waiting on standard tools
WOT	Waiting on test or tools
WOT&C	Waiting on tank & connection
WOW	Waiting on weather
WP	Wash pipe, working pressure
wpr	Wrapper
WR	White River
Wref	Wreford
WS	Whipstock

WSD	Whipstock depth
w shd	Washed
wshg	Washing
WSO	Water shut-off
WSONG	Water shut off no good
WSOOK	Water shut off OK
W/SSO	Water with slight show of oil
W/sulf O	Water with sulphur odor
WSW	Water supply well
WT	Wall thickness (pipe)
wt	Weight
wtg	Waiting
wthd	Weathered
wthr	Weather
wtr(y)	Water, watery
WTS	Water to surface
WW	Wash water, water well
Wx	Wilcox

X

X-bdd(ing)	Crossbedded, crossbedding
X-hvy	Extra heavy
Xing	Crossing
Xlam	Cross-laminated
X-line	Extreme line (casing)
Xln	Crystalline
X-over	Crossover
X-R	X-ray
x-stg	Extra strong
xtal	Crystal
Xtree	Christmas tree
XX-Hvy	Double extra heavy

Y

Y	Yates
yd	Yard(s)
YIL	Yellow indicating lamp
yel	Yellow
YMD	Your message of date
YMY	Your message yesterday
Yoak	Yoakum
YP	Yield point
yr	Year
Yz	Yazoo

Z

Z	Zone
zen	Zenith
Zil	Zilpha

A

Abandoned	abd
Abandoned gas well	abd-gw
Abandoned location	abd loc
Abandoned oil & gas well	abdogw
Abandoned oil well	abd-ow
About	abt
Above	ab
Above	abv
Abrasive jet	abrsi jet
Absolute open flow potential (gas well)	AOF
Absorber	asbr
Absorption	absrn
Abstract	abst
Abundant	abun
Account	acct
Accounting	acct
Account of	acct
Accounts receivable	A/R
Accumulative	accum
Acid	ac
Acid-cut mud	ACM
Acid-cut water	ACW
Acid frac	AF
Acid fracture treatment	acfr
Acidize	acd
Acidized	acd
Acidized with	A/
Acidizing	ac
Acidizing	acd
Acid residue	AR
Acid treat(ment)	AT
Acid water	AW
Acoustic cement	A-Cem
Acre	ac

Acreage	ac
Acreage	acrg
Acre feet	ac-ft
Acres	ac
Acrylonitrile butadiene styrene rubber	ABS
Adapter	adpt
Additional	addl
Additive	add
Adjustable	adj
Adjustments and Allowances	A&A
Administration	adm
Administrative	adm
Adomite	ADOM
Adsorption	adspn
Advanced	advan
Affidavit	afft
Affirmed	affd
After acidizing	AA
After fracture	AF
After shot	AS
After top center	ATC
After treatment	AT
Agglomerate	aglm
Aggregate	aggr
Alarm	alm
Albany	Alb
Algae	alg
Alkalinity	alk
Alkylate	alky
Alkylation	alky
Allowable	allow
Allowable not yet available	ANYA
All thread	AT
Alternate	alt
Alternating current	AC
Aluminum conductor steel reinforced	ACSR
Ambient	amb
American Steel & Wire gauge	AS&W ga
American Wire Gauge	AWG
Amorphous	amor

Amortization	amort
Amount	amt
Ampere	amp
Ampere hour	amp hr
Amphipore	amph
Amphistegina	Amph
Analysis	anal
Analytical	anal
And husband	et con
And husband	et vir
And the following	et seq
And others	et al
And wife	et ux
Angle	ang
Angular	ang
Angulogerina	Angul
Anhydrite	anhy
Anhydrite stringer	AS
Anhydritic	anhy
Anhydrous	anhyd
Annular velocity	AV
Apartment	apt
Apparent(ly)	apr
Appearance	app
Appears	app
Appliance	appl
Application	applic
Applied	appl
Approved	appd
Approximate(ly)	approx
Aqueous	aq
Aragonite	arag
Arapahoe	Ara
Arbuckle	Arb
Archeozoic	Archeo
Architectural	arch
Arenaceous	aren
Argillaceous	arg
Argillite	arg
Arkadelphia	Arka

Arkose(ic)	ark
Armature	arm
Aromatics	arom
Around	arnd
As above	AA
Asbestos	asb
Ashern	Ash
Asphalt	asph
Asphaltic	asph
Asphaltic stain	astn
Assembly	assy
Assigned	assgd
Assignment	asgmt
Assistant	asst
Associate(d)(s)	assoc
Association	assn
As soon as possible	ASAP
Atmosphere	atm
Atmospheric	atm
Atoka	At
Atomic	at
Atomic weight	at wt
At rate of	ARO
Attempt(ed)	att
Attorney	atty
Audit Bureau of Circulation	ABC
Auditorium	aud
Austin	Aus
Austin Chalk	AC
Authorization for expenditure	AFE
Authorized	auth
Authorized depth	AD
Automatic	auto
Automatic custody transfer	ACT
Automatic data processing	ADP
Automatic transmission fluid	ATF
Automatic volume control	AVC
Automotive	auto
Automotive gasoline	autogas
Aux Vases sand	AV

Auxiliary	aux
Available	avail
Average	avg
Average flowing pressure	AFP
Average injection rate	AIR
Aviation	av
Aviation gasoline	avgas
Awaiting	awtg
Azeotropic	aztrop
Azimuth	az

B

Back pressure	BP
Back pressure valve	BPV
Back to back	B/B
Backed out (off)	BO
Bailed	bld
Bailed dry	B/dry
Bailed water	BW
Bailer	blr
Bailing	blg
Ball and flange	B&F
Balltown sand	Ball.
Band(ed)	bnd
Barge deck to mean low water	BD-MLW
Barite(ic)	bar
Barker Creek	Bark Crk
Barlow Lime	Bar
Barometer	bar
Barometric	bar
Barrel	BBL
Barrels	BBL
Barrels acid residue	BAR
Barrels acid water	BAW
Barrels acid Water Per Day	BAWPD
Barrels Acid Water Per Hour	BAWPH

Barrels acid water under load	BAWUL
Barrels condensate per day	BCPD
Barrels condensate per hour	BCPH
Barrels condensate per million	BCPMM
Barrels diesel oil	BDO
Barrels distillate per day	BDPD
Barrels distillate per hour	BDPH
Barrels fluid	BF
Barrels fluid per day	BFPD
Barrels fluid per hour	BFPH
Barrels formation water	BFW
Barrels frac oil	BFO
Barrels load	BL
Barrels load & acid water	BL&AW
Barrels load oil	BLO
Barrels load oil recovered	BLOR
Barrels load oil yet to recover	BLOYR
Barrels load water	BLW
Barrels mud	BM
Barrels new oil	BNO
Barrels of acid	BA
Barrels of condensate	BC
Barrels of distillate	BD
Barrels of pipeline oil	BPLO
Barrels of pipeline oil per day	BPLOPD
Barrels of water	BW
Barrels of water per day	BW/D
Barrels of water per day	BWPD
Barrels of water per hour	BWPH
Barrels oil	BO
Barrels oil per calendar day	BOCD, BOPCD
Barrels oil per day	BOD
Barrels oil per day	BOPD
Barrels oil per hour	BOPH
Barrels oil per producing day	BOPPD
Barrels per barrel	B/B
Barrels per calender day	BPCD
Barrels per day	B/D
Barrels per hour	B/H

Barrels per hour	B/hr
Barrels per hour	BPH
Barrels per minute	BPM
Barrels per stream day (refinery)	B/SD
Barrels per well per day	BPWPD
Barrels salt water	BSW
Barrels salt water per day	BSWPD
Barrels salt water per hour	BSWPH
Barrels water load	BWL
Barrels water over load	BWOL
Bartlesville	Bart
Basal Oil Creek sand	BOCS
Base	B/
Base Blane	B.Bl
Base of the salt	B slt, B/S
Base Pennsylvanian	BP
Base plate	BSPL
Basement	bsmt
Basement (Granite)	base
Basic sediment	BS
Basic sediment & water	BS&W
Basket	bskt
Bateman	Bate
Battery	bat
Battery	btry
Baume	Be
Beaded and center beaded	B & CB
Bearing	brg
Bearpaw	BP
Bear River	Bear R
Becoming	bec
Beckwith	Beck
Before acid treatment	BAT
Before top dead center	BTDC
Beldon	Bel
Belemnites	Belm
Bell and bell	B&B
Bell and spigot	B&S
Belle City	Bel C
Belle Fourche	Bel F

Bench mark	BM
Benoist (Bethel) sand	Ben, BT
Benton	Ben
Bentonite	bent
Bentonitic	bent
Benzene	bnz
Benzene toluene, xylene (Unit)	BTX(unit)
Berea	Be
Bethel (Benoist) sand	BT, Ben
Between	btw
Bevel	bev
Bevel both ends	BBE
Beveled as for welding	bev
Beveled for welding	BV/WLD
Beveled end	B.E.
Bevel one end	BOE
Bevel small end	BSE
Big Horn	B Hn
Big Injun	B. Inj.
Big Lime	B. Ls
Bigenerina	Big.
Bigenerina floridana	Big. f.
Bigenerina humblei	Big. h.
Bigenerina nodosaria	Big. nod.
Billion cubic feet	BCF
Billion cubic feet per day	BCFD
Bill of lading	B/L
Bill of material	B/M
Bill of sale	B/S
Biotite	bio
Birmingham (or Stubbs) iron wire gauge	BW ga
Bitumen	bit
Bituminous	bit
Black	blk
Black Leaf	Blk Lf
Black Lime	Blk Li
Black Magic (mud)	BM
Black malleable iron	BMI
Black River	B. Riv
Black sulfur water	BSUW

Blank liner	blk lnr
Blast joint	BL/JT
Bleeding	bldg
Bleeding gas	bldg
Bleeding oil	bldo
Blend	blnd
Blended	blnd
Blender	blndr
Blending	blnd
Blind (flange)	bld
Blinebry	Blin
Block	blk
Blossom	Blos
Blow	blo
Blow-down test	BDT
Blow out equipment	BOE
Blowout preventer	BOP
Blue	bl
Board	bd
Board feet	bd ft
Board foot	bd ft
Bodcaw	Bod
Boiler feed water	BFW
Boiling point	BP
Bois D'Arc	Bd'A
Bolivarensis	Bol.
Bolivina a.	Bol. a.
Bolivina floridana	Bol. flor.
Bolivina perca	Bol. p.
Bone Spring	BS
Bonneterre	Bonne
Bottom	bot
Bottom	btm
Bottom choke	BC
Bottom choke	btm chk
Bottomed	btmd
Bottom-hole choke	BHC
Bottom-hole flowing pressure	BHFP
Bottom-hole location	BHL
Bottom-hole money	BHM

Bottom-hole pressure	BHP
Bottom-hole pressure, closed (See also SIBHP and BHSIP)	BHPC
Bottom-hole pressure, flowing	BHPF
Bottom-hole pressure survey	BHPS
Bottom-hole shut-in pressure	BHSIP
Bottom-hole assembly	BHA
Bottom-hole temperature	BHT
Bottom of given formation (i.e., B/Frio)	B/
Bottom sediment	BS
Bottom settlings	BS
Boulders	bldrs
Boundary	bndry
Box(es)	bx
Brachiopod	brach
Bracket(s)	brkt(s)
Brackish (water)	brksh
Brake horsepower	bhp
Brake horsepower hour	bhp-hr
Brake mean effective pressure	BMEP
Brake specific fuel consumption	BSFC
Break (broke)	brk
Breakdown	bkdn
Breakdown acid	BDA
Breakdown pressure	BDP
Breaker	bkr
Breccia	brec
Bridged back	BB
Bridge plug	BP
Bridger	Brid
Brinell hardness number	BHN
British thermal unit	BTU
Brittle	brit
Brittle	brtl
Broke (break) down formation	BDF
Broken	brkn
Broken sand	brkn sd
Bromide	Brom
Brown	bn
Brown	brn or br

Brown and Sharpe gauge	B&S ga
Brown lime	Brn Li
Brown oil stain	BOS
Brown shale	brn sh
Brownish	bnish
Bryozoa	bry
Buckner	Buck
Buckrange	Buckr
Budgeted depth	BD
Buff	bf
Building	bldg.
Building derrick	bldg drk
Building rig	BR
Building road	BR
Building roads	bldg rds
Buliminella textularia	Bul. text.
Bulk plant	BP
Bullets	blts
Bull plug	BP
Bull waggon	Bull W
Bumper	bmpr
Burgess	Burg
Burner	bunr
Bushing	bsg
Butane	BB fraction
Butane and propane mix	BP Mix
Butene fraction	BB fraction
Butt weld	BW
Buttress thread	butt

C

Cable tools	CT
Caddell	Cadd
Calcareous	calc
Calceneous	cale
Calcerenite	calc

Calcite	cal
Calcitic	cal
Calcium	calc
Calcium base grease	calc gr
Calculate(d)	calc
Calculated absolute open flow	CAOF
Calculated open flow	COF
Calculated open flow (potential)	calc OF, COF
Calendar day	CD
Caliche	cal
Caliper survey	cal
Calorie	cal
Calvin	Calv
Cambrian	Camb
Camerina	Cam.
Camp Colorado	Cp Colo
Cane River	CR
Cane River	Cane R
Canvas-lined metal petal basket	CLMP
Canyon	Cany
Canyon Creek	Cany Crk
Capacitor	cap
Capacity	cap
Capitan	Cap
Carbonaceous	carb
Carbon copy	CC
Carbon Residue (Conradson)	CR Con
Carbon steel	CS
Carbon tetrachloride	carb tet
Carburetor air temperature	CAT
Care of	c/o
Carlile	Car
Carload	CL
Carmel	Carm
Carrizo	Cz
Carton	ctn
Cased Hole	C/H
Casing	csg
Casing cemented (depth)	CC
Casing choke	Cck

Casing collar locator	CCL
Casing collar perforating record	CCPR
Casing flange	CF
Casing head	csg hd
Casinghead flange	CHF
Casinghead (gas)	CH
Casinghead gas	CHG
Casinghead pressure	CHP
Casing point	CP
Casing point	csg pt
Casing pressure	CP
Casing pressure	csg press
Casing pressure–closed	CPC
Casing pressure–flowing	CPF
Casing pressure shut in	CPSI
Casing seat	CS
Casing set at	CSA
Casper	Casp
Cast-iron	CI
Cast-iron bridge plug	CIBP
Cast-steel	CS
Catcracked light gas oil	CCLGO
Cat Creek	Cat Crk
Catahoula	Cat
Catalog	cat
Catalyst	cat
Catalytic	cat
Catalytic cracker	cat ckr
Catalytic Cracking Unit	CCU
Cathodic	cath
Cattleman	Ctlmn
Caustic	caus
Caving(s)	cvg(s)
Cavity	cav
Cedar Mountain	Cdr Mtn
Cellar	cell
Cellar & pits	C&P
Cellular	cell
Cement(ed)	cem
Cement(ed) (ing)	cmt(d) (g)

Cement friction reducer	CFR
Cement in place	CIP
Cementer	cmtr
Cenozoic	Ceno
Center	ctr
Center(ed)	cntr
Center (land description)	C
Center line	C/L
Center of casinghead flange	CCHF
Center of gravity	CG
Center of tubing flange	CTHF
Center to center	C to C
Center to end	C to E
Center to face	C to F
Centigrade temp. scale	C
Centimeter	cm
Centimeter-gram-second system	cgs
Centimeters per second	cm/sec
Centipoise	cp
Centistokes	cs
Centralizers	cent
Centrifugal	centr
Centrifuge	cntf
Cephalopod	ceph
Ceratobulimina eximia	Cert
Certified public accountant	CPA
Cetane number	CN
Chairman	chrm
Chalcedony	chal
Chalk	chk
Chalky	chky
Change	chng
Changed	chng
Changed(ing) bits	CB
Changing	chng
Chappel	Chapp
Charge	chg
Charged	chg
Charging	chg
Charles	Char

Chart	cht
Chattanooga shale	Chatt
Check	ck
Checked	chkd
Checkerboard	Chkbd
Chemical	chem
Chemically pure	CP
Chemically retarded acid	CRA
Chemical products	chem prod
Chemist	chem
Chemistry	chem
Cherokee	Cher
Chert	ch
Chert	cht
Cherty	chty
Chester	Ches
Chicksan	cksn
Chimney Hill	Chim H
Chimney Rock	Chim R
Chinle	Chin
Chitin(ous)	chit
Chloride(s)	chl
Chlorine log	chl log
Chloritic	chl
Choke	ch
Choke	chk
Chouteau lime	Chou
Christmas tree	Xtree
Chromatograph	chromat
Chrome molybdenum	cr moly
Chromium	chrome
Chugwater	Chug
Cibicides	Cib
Cibicides hazzardi	Cib h
Cimarron	Cima
Circle	cir
Circuit	cir
Circular	cir
Circular mils	cir mils
Circulate	circ

Circulated out	CO
Circulating	circ
Circulating & conditioning	C&C
Circulation	circ
Cisco	Cis
Clagget	Clag
Claiborne	Claib
Clarksville	Clarks
Clastic	clas
Clavalinoides	Clav
Clay filled	CF
Claystone	clyst
Clayton	Clay
Claytonville	Clay
Clean, cleaned, cleaning	cln (d) (g)
Cleaned out	CO
Cleaned out to total depth	COTD
Cleaning out	CO
Cleaning to pits	CTP
Clean out	CO
Clean out & shoot	CO & S
Clean up	CU
Clear	clr
Clearance	clr
Clearfork	Clfk
Clearing	clrg
Cleveland	Cleve
Cleveland open cup	COC
Cliff House	Cliff H
Clockwise	cw
Closed	clsd
Closed-in pressure	CIP
Closed cup	CC
Cloverly	Clov
Coarse-grained	c-gr
Coarse-grained	cse gr
Coarse(ly)	c
Coarse(ly)	crs
Coat and wrap (pipe)	C & W
Coated	ctd

Coated and wrapped	ctw
Cockfield	Cf
Coconino	Coco
Codell	Cod
Cody (Wyoming)	Cdy
Coefficient	coef
Coke oven gas	COG
Cold rolled	CR
Cold working pressure	CWP
Cole Junction	Cole J
Coleman Junction	Col Jct
Collar	colr
Collect	coll
Collected	coll
Collecting	coll
Collection	coll
Color, American Standard Test Method	Col ASTM
Colored	COL
Column	COL
Comanche	Com
Comanchean	Cmchn
Comanche Peak	Com Pk
Comatula	Com
Combination	comb
Combined	comb
Coming out of hole	COH
Commenced	comm
Commercial	coml
Commission	comm
Commission agent	C/A
Commissioner	commr
Common	com
Common Business Oriented Language	COBOL
Communication	comm
Communitized	comm
Community	comm
Compact	cmpt
Companion flange one end	CFOE
Companion flanges bolted on	CFBO
Company operated	Co. Op.

Company operated service stations	Co. Op. S.S.
Compartment	compt
Complete	comp
Completed	compt
Completed natural	comp nat
Completed with	C/W
Completion	comp
Components	compnts
Compressor	compr
Compressor station	compr sta
Compression and absorption plant	C&A
Compression-ignition engine	CI engine
Compression ratio	CR
Concentrate	conc
Concentric	cnnc
Concentric	conc
Conchoidal	conch
Conclusion	concl
Concrete	conc
Concretion(ary)	conc
Condensate	cond
Condensate-cut mud	CCM
Condensor	cdsr
Conditioned	cond
Conditioning	cond
Conductivity	condt
Conductor (pipe)	condr
Confirm	conf
Confirmed	conf
Confirming	conf
Conflict	confl
Conglomerate	cglt
Conglomerate(itic)	cong
Conglomeritic	cglt
Connection	conn
Conodonts	cono
Conradson carbon residue	CCR
Conservation	consv
Conserve	consv
Consolidated	con

Consolidated	consol
Constant	const
Construction	const
Consumer tank car	CTC
Consumer tank wagon	CTW
Consumer transport truck	CTT
Contact caliper	C-Cal
Container	cntr
Contaminated	contam
Contamination	contam
Continue, continued	cont(d)
Continuous dipmeter survey	CDM
Continuous weld	CW
Contour interval (map)	CI
Contract depth	CD
Contractor (i.e., C/John Doe)	C/
Contractor	contr
Contractor's responsibility	contr resp
Contribution	contrib
Controller	cntr
Control(s)	cntl
Control valve	CV
Converse	conv
Conveyor	cnvr
Coquina	coq
Cook Mountain	Ck Mt
Cooperative	co-op
Coordinate	coord
Coordinating Research Council, Inc.	CRC
Core, cored, coring, core hole	cr (d), (g), (h)
Core barrel	CB
Coring	cg
Corner	cor
Corporation	Corp
Corrected gravity	CG
Correct (ed) (ion), corrosion, corrugated	corr
Corrected total depth	CTD
Correlation	correl
Correspondence	corres
Cost, insurance and freight	CIF

Cost per gallon or cents per gallon	CPG
Cottage Grove	Cott G
Cotton Valley	CV
Cottonwood	Ctnwd
Council Grove	Counc G
Counterbalance (pumping equip)	CB
Counterclockwise	ccw
Counter electromotive force	CEMF
County	Cnty
County school lands, center section line	CSL
Coupling	cplg
Cow Run	CR
Cracker	crkr
Cracking	crkg
Crenulated	cren
Cretaceous	Cret
Crinkled	crnk
Crinoid(al)	Crin
Cristellaria	Cris
Critical	crit
Critical compression pressure	CCP
Critical compression ratio	CCR
Cromwell	Crom
Crossbedded	Crbd
Crossbedded	X-bdd(ing)
Crossbedding	X-bdd(ing)
Crossing	Xing
Cross-laminated	Xlam
Crossover	CX
Crossover	X-over
Crown block	crn blk
Crude oil	CO
Crude oil purchasing	COP
Cryptocrystalline	crypto-xln
Crystal	xtal
Crystalline	cryst
Crystalline	Xln
Cubic	cu
Cubic centimeter	cc
Cubic feet	CF

Cubic feet gas	CFG
Cubic feet gas per day	CFGPD
Cubic feet of gas per hour	CFGH
Cubic feet per barrel	cu ft/bbl
Cubic feet per minute	CFM
Cubic feet per minute	cu ft/min
Cubic feet per second	CFS
Cubic feet per second	cu ft/sec
Cubic foot	cu ft
Cubic inch	cu in
Cubic meter	cu m
Cubic yard	cu yd
Culvert	culv
Cumulative	cum
Curtis	Cur
Cushion	cush
Cut across grain	CAG
Cutback	cutbk
Cut Bank	Cut B
Cutler	Cutl
Cutting oil	Cut Oil
Cutting oil-active sulphurized-dark	Cut Oil Act Sul-Dk
Cutting oil-active-sulphurized transparent	Cut Oil Act Sul-Trans
Cutting oil-inactive-sulphurized	Cut Oil Inact Sul
Cutting oil soluble	Cut Oil Sol
Cutting oil–straight mineral	Cut Oil St Mrl
Cuttings	ctg(s)
Cyclamina	Cyc.
Cyclamina cancellata	Cycl canc
Cycles per minute	CPM
Cycles per second	CPS
Cylinder	cyl
Cylinder Stock	cyl stk
Cypress sand	Cy Sd
Cypridopsis	cyp

D

Daily allowable	DA
Daily average injection, barrels	DAIB
Dakota	Dak
Damper	dmpr
Dantzler	Dan
Dark	dk
Dark brown oil	DBO
Dark brown oil stains	DBOS
Darwin	Dar
Data processing	DP
Date of first production	DFP
Datum	dat
Datum	DM
Datum faulted out	DFO
Days since spudded	DSS
Day to day	D/D
Dead	dd
Dead oil show	DSO
Dead weight tester	DWT
Deadwood	Deadw
Deaerator	deaer
Dealer	dlr
Dealer tank wagon	DTW
Deasphalting	deasph
Debutanizer	debutzr
Decibel	db
Decimal	dec
Decrease (d) (ing)	decr
Deepen	dpn
Deepening	dpg
Deep pool test	DPT
Deethanizer	deethnzr
Deflection	defl
Degonia	deg
Degree day	DD
Degree Fahrenheit	°F.

Deisobutanizer	deisobut
Delaware	Dela
Delayed coker	DC
Deliverability	delv
Delivered	delv
Delivery	delv
Delivery point	delv pt
Del Rio	Del R
Demand meter	DM
Demolition	dml
Demurrage	demur
Dendrite(ic)	dend
Dense	dns
Dense	ds
Density log	D/L
Density log	DENL
Department	dept
Depletion	depl
Depreciation	dep
Depreciation	deprec
Depropanizer	deprop
Depth	dpt
Depth recorder	dpt rec
Derrick	drk
Derrick floor	DF
Derrick floor elevation	DFE
Desalter	desalt
Description	desc
Desert Creek	Des Crk
Design	dsgn
Desk and Derrick	D & D
Des Moines	Des M
Desorbent	desorb
Destination	dstn
Desulferizer	desulf
Detail(s)	det
Detector	det
Detergent	deterg
Detrital	detr
Detrital	dtr

Develop (ed) (ment)	devel
Development gas well	DG
Development oil well	DO
Development redrill (sidetrack)	DR
Development well–carbon dioxide	DC
Development well–helium	DH
Development well–sulfur	DSU
Development well workover	DX
Deviate	dev
Deviation	dev
Devonian	Dev
Dewaxing	dewax
Dew point	DP
Dexter	Dext
Diagonal	diag
Diagram	diag
Diameter	dia
Diamond core	DC
Diamond core bit	DCB
Diaphragm	diaph
Dichloride	dichlor
Dichloro-diphenyl-trichloroethane	DDT
Diesel No. 2	D-2
Diesel fuel	DF
Diesel (oil)	dsl
Diesel oil cement	DOC
Diethylene	diethy
Difference	diff
Different	diff
Differential	diff
Differential valve	DV
Digging slush pits, digging cellar, or digging cellar and slush pits	DCLSP
Diluted	dilut
Dimension	dim
Diminish	dim
Diminishing	dim
Dinwoody	Din
Dip meter	DM
Direct	dir

Direct current	DC
Direction	dir
Directional survey	DS
Directional survey	dir sur
Director	dir
Discharge	disch
Discorbis	Disc
Discorbis gravelli	Disc grav
Discorbis normada	Disc norm
Discorbis yeguaensis	Disc y
Discount	disc
Discovery allowable requested	DAR
Discover (y) (ed) (ing)	disc
Disseminated	dism
Dismantle	disman
Dismantle (ing)	dismtl (g)
Displaced	displ
Displacement	displ
Distance	dist
Distillate	dist
Distillate	dstl
Distillate-cut mud	DCM
Distillation	dist
Distribute (d) (ing) (ion)	distr
District	dist
Ditto	do
Division	div
Division office	D/O
Division order	D.O.
Dockum	Doc
Doctor-treating	doc-tr
Document	doc
Doing business as	d/b/a
Dolomite(ic)	dolo
Dolstone	dolst
Domestic	dom
Domestic airline	dom AL
Dornick Hills	Dorn H
Dothan	Doth
Double end	DE

Double extra heavy	XX-Hvy
Double pole double base (switch)	DPDB
Double pole double throw (switch)	DPDT
Double pole single base (switch)	DPSB
Double pole single throw (switch)	DPST
Douple pole (switch)	DP
Double random lengths	DRL
Douglas	Doug
Down	dn
Down	dwn
Dozen	doz
Draft gage	DG
Drain	dr
Drainage	drng
Drawing	dwg
Drawworks	dwks
Dressed dimension four sides	d-d-4-s
Dressed dimension one side and one edge	d-d-1-s-1-e
Dressed four sides	d-4-s
Dressed one side	d-1-s
Dressed two sides	d-2-s
Drier	dry
Drill	drl
Drill and complete	D & C
Drill collar	DC
Drill collars	DCS
Drilled	drld
Drill (ed) (ing) out	DO
Drill (ed) (ing) plug	D/P
Drilled out cement	DOC
Drilled out depth	DOD
Drilled out plug	DOP
Driller	drlr
Driller's tops	D/T
Drillers total depth	DTD
Drilling	drlg
Drilling break	DB
Drilling deeper	DD
Drilling mud	DM
Drilling suspended indefinitely	DSI

Drilling time	DT
Drilling with air	DWA
Drilling with mud	DWM
Drilling with oil	DWO
Drilling with salt water	DWSW
Drill pipe	DP
Drill pipe measurement	DPM
Drill pipe unloaded	DPU
Drill stem	DS
Drill stem test	DST
Drive	dr
Dropped	drpd
Drum	dr
Druse	dr
Drusy	drsy
Dry and abandoned	D&A
Dry Creek	Dr Crk
Dry desiccant dehydrator	DDD
Dry gas	DG
Dry hole contribution	DHC
Dry hole drilled deeper	DHDD
Dry hole money	DHM
Dry hole reentered	DHR
Drying	dry
Dually complete(d)	DC
Dual (double) wall packer	DWP
Duck Creek	Dk Crk
Dun & Bradstreet	D & B
Duperow	Dup
Duplex	dx
Duplicate	dup
Dutcher	Dutch
Dynamic	dyn

E

Each	ea
Eagle	Egl
Eagle Ford	EF
Eagle Mills	EM
Earlsboro	Earls
Earthy	rthy
East	E
East boundary line	E/BL
East Cimarron Meridian (Oklahoma)	ECM
East half, quarter, etc.	E/2, E/4
East Line	E/L
East offset	E/O
East of Rockies	EOR
East of west line	E of W/L
Eau Claire	Eau Clr
Eccentric	ecc
Echinoid	Ech
Economics	Econ
Economizer	Econ
Economy	Econ
Ector (county, Tex.)	Ect
Education	educ
Edwards	Edw
Edwards lime	Ed lm
Effective	eff
Effective horsepower	EHP
Efficiency	eff
Effluent	effl
Ejector	eject
Elbert	Elb
Elbow(s)	ell(s)
Electric accounting machines	EAM
Electric(al)	elec
Electric log tops	El/T
Electric resistance weld	ERW
Electric weld	EW

Electromotive force	EMF
Electronic data processing	EDP
Electron volts	ev
Element(ary)	elem
Elevation	elev
Elevation ground	el gr
Elevator	elev
Elgin	Elg
Ellenburger	Ellen
Ellis-Madison contact	EMS
Elmont	Elm
Embar	Emb
Emergency	emer
Emergency order	EO
Employee	empl
Empty container	MT
Emulsion	emul
Enamel	enml
Enclosure	encl
Endicott	End
End of file	EOF
End of line	EOL
End of month	EOM
End of quarter	EOQ
End of year	EOY
Endothyra	endo
End point	EP
End to end	E/E
Engine	eng
Engineer(ing)	engr(g)
Englewood	Eglwd
Enlarged	enl
Entrada	Ent
Eocene	Eoc
Eponides	Epon
Eponides yeguaensis	Ep y
Equal	eq
Equalizer	eq
Equation	eq
Equilibrium flash vaporization	EFV

Equipment	equip
Equivalent	equiv
Erection	erect
Ericson	Eric
Estate	est
Estimate (d) (ing)	est
Estimated time of arrival	ETA
Estimated yearly consumption	EYC
Ethane	eth
Ethylene	ethyle
Euhedral	euhed
Eutaw	Eu
Evaluate	eval
Evaporation	evap
Evaporite	evap
Even-sorted	ev-sort
Excavation	exc
Excellent	excl
Exchanger	exch
Executor	exr
Executrix	exrx
Exhaust	exh
Exhibit	exh
Existing	exist
Existing	exst
Expansion	exp
Expendable plug	exp plg
Expense	exp
Expiration	expir
Expire	expir
Expired	expir
Expiring	expir
Exploration	expl
Exploratory	expl
Exploratory well	EW
Explosive	explos
Extended	ext(n)
Extension	ext(n)
Extension manhole	Ext M/H
Exter	Ex

Exterior	extr
External	ext
External upset end	EUE
Extraction	extrac
Extra heavy	X-hvy
Extra strong	x-stg
Extreme pressure	EP
Extreme line (casing)	X-line

F

Fabricate(d)	fab
Faced and drilled	F&D
Facet(ed)	fac
Face to face	F to F
Failed	fld
Failure	fail
Faint	fnt
Faint air blow	FAB
Fair	fr
Fall River	Fall Riv
Farmington	Farm
Farmout	FO
Fault	flt
Faulted out	FO
Fauna	fau
Favosites	fvst
Federal	fed
Federal Employers Liability Act	FELA
Feed	FD
Feeder	fdr
Feed rate	FR
Feedstock	FS
Feet	ft
Feet per hour	ft/hr
Feet per minute	fpm
Feet per minute	ft/min

Feet per second	fps
Feet per second	ft/sec
Feldspar (rhic)	fld
Female pipe thread	FPT
Female to female angle	FFA
Female to female globe (valve)	FFG
Ferguson	Ferg
Ferruginous	ferr
Ferry Lake anhydrite	FLA
Fertilizer	fert
Fiberglass reinforced plastic	FRP
Fibrous	fib
Field	fld
Field authorized to commence operations	FACO
Field purchase order	FPO
Field wildcat	FWC
Figure	fig
Fillister	fill
Fill up	FU
Filter cage	FC
Filtrate	filt
Final	fin
Final boiling point	FBP
Final bottom-hole pressure flowing	FBHPF
Final bottom-hole pressure shut-in	FBHPSI
Final flowing pressure	FFP
Final hydrostatic pressure	FHP
Final pressure	FP
Final report for rig	FRR
Final report for well	FRQ
Final shut-in pressure	FSIP
Final tubing pressure	FTP
Fine	fn
Fine-grained	f-gr
Finely	fnly
Finely-crystalline	f/xln
Finish	fin
Finish all over	FAO
Finished	fin
Finished drilling	fin drlg

Finish going in hole	FGIH
Finish going in with—	FGIW
Fireproof	fprf
Fire-resistant oil	F-R Oil
Fishing	fish
Fishing	fsg
Fishing for	FF
Fissile	fisl
Fissure	fis
Fittings	ftg
Fixed	fxd
Fixed carbon	FC
Fixture	fix
Flaky	flk
Flanged and dished (heads)	F & D
Flanged and spigot	F & S
Flanged one end, welded one end	FOE—WOE
Flange (d) (s)	flg (d) (s)
Flash Point, Cleveland Open Cup	Fl–COC
Flat face	FF
Flathead	Flath
Flattened	flat
Flexible	flex
Flippen	Flip
Float	flt
Float collar	FC
Floating	fltg
Float shoe	FS
Floor	FL
Floor drain	FD
Florence Flint	Flor Fl
Flow	flo
Flow control valve	FCV
Flow indicator	FI
Flow indicating controller	FIC
Flow indicating ratio controller	FIRC
Flow line	FL
Flow rate	FR
Flow recorder	FR
Flow recorder control	FRC

Flowed or flowing	fl/
Flowed, flowing	F/
Flowed, flowing	flw (d) (g)
Flowed (ing) at rate of	FARO
Flowerpot	Flwrpt
Flowing	flg
Flowing bottom-hole pressure	FBHP
Flowing by heads	FBH
Flowing casing pressure	FCP
Flowing on test	FOT
Flowing pressure	Flwg. Pr.
Flowing pressure	FP
Flowing surface pressure	FSP
Flowing tubing pressure	FTP
Flue	flu
Fluid in hole	FIH
Fluid	fl
Fluid	flu
Fluid catalytic cracking	FCC
Fluid level	FL
Fluid to surface	FTS
Fluorescence	fluor
Fluorescent	fluor
Flush	FL
Flushed	flshd
Flush joint	FJ
Focused log	FOCL
Foliated	fol
Foraker	forak
Foraminifera	foram
Foreman	f'man
For example	e.g.
For your information	FYI
Forged steel	FS
Forged steel	FST
Formation	fm
Formation density	F-D
Formation density log	FDL
Formation gas-oil ratio	F/GOR
Formation interval tester	FIT

Formation water	Fm W
Fort Chadbourne	Ft C
Fort Hayes	Ft H
Fort Riley	Ft R
Fort Worth	Ft W
Fort Union	Ft U
Fortura	Fort
Forward	fwd
Foot	ft
Footage	ftg
Foot-candle	ft-c
Footing	ftg
Foot-pound	ft lb
Foot-pound-second (system)	fps
Foot pounds per hour	ft lbs/hr
Fossiliferous	foss
Foundation	fdn
Fountain	Fount
Four-wheel drive	FWD
Fox Hills	Fox H
Frac finder (Log)	FF
Fractional frosted	fr
Fractionation	fract
Fractionator	fract
Fracture, fractured, fractures	frac (d) (s)
Fracture gradient	F.G.
Fragment	frag
Framework	frwk
Franchise	fran
Franconia	Franc
Fredericksburg	Fred
Fredonia	Fred
Free on board	FOB
Free point indicator	FPI
Freezer	frzr
Freezing point	FP.
Freight	frt
Frequency	freq
Frequency meter	FM
Frequency modulation	FM

Fresh	frs
Fresh break	FB
Fresh water	FW
Friable	fri
Friction reducing agent	FRA
Froggy	Frgy
From	fr
From east line	FE/L
From east line	fr E/L
From north line	FNL
From north line	fr N/L
From northeast line	FNEL
From northwest line	FNWL
From south and west lines	FS&WLs
From southeast line	FSEL
From south line	fr S/L
From south line	FSL
From southwest line	FSWL
From west line	fr W/L
From west line	FWL
Front	fr
Front & side, flange x screwed	F/S
Frontier	Fron
Frosted	fros
Frosted quartz grains	FQG
Fruitland	Fruit
Fuel oil	FO
Fuel oil equivalent	F.O.E.
Fuels & fractionation	F&F
Fuels & Lubricants	F & L
Fullerton	Full
Full hole	FH
Full interest	FI
Full of fluid	FF
Full open head (grease drum 120 lb)	FOH
Full opening	FO
Funnel viscosity	FV
Furfural	furf
Furnace	furn
Furnace Fuel Oil	FFO

Furniture and fixtures	Furn & fix
Fuson	Fus
Fusselman	Fussel
Future	fut
Fusulinid	Fusul

G

Gage (d) (ing)	ga
Galena	Glna
Gallatin	Gall
Gallon	gal
Gallons	gal
Gallons acid	GA
Gallons breakdown acid	GBDA
Gallons condensate per day	GCPD
Gallons condensate per hour	GCPH
Gallons gelled water	GGW
Gallons heavy oil	GHO
Gallons mud acid	GMA
Gallons oil	GO
Gallons oil per hour	GOPH
Gallons oil per day	GOPD
Gallons solution	gal sol
Gallons water per hour	GWPH
Gallons per day	GPD
Gallons per hour	GPH
Gallons per minute	gal/min
Gallons per minute	GPM
Gallons per thousand cubic feet	gal/Mcf
Gallons per thousand cubic feet	gpm
Gallons per second	GPS
Gallons regular acid	GRA
Gallons salt water	GSW
Gallons water	GW
Galvanized	galv
Gamma ray	GR

Gas	G
Gas and mud-cut oil	G&MCO
Gas and oil	G&O
Gas and oil-cut mud	G&OCM
Gas and water	G&W
Gas-condensate ratio	GCR
Gas cut	GC
Gas-cut acid water	GCAW
Gas-cut distillate	GCD
Gas-cut load oil	GCLO
Gas-cut load water	GCLW
Gas-cut mud	GCM
Gas-cut oil	GCO
Gas-cut salt water	GCSW
Gas-cut water	GCW
Gas-distillate ratio	GDR
Gas injection	GI
Gas-injection well	GIW
Gasket	gskt
Gas lift	GL
Gas-liquid ratio	GLR
Gas odor	GO
Gas odor distillate taste	GODT
Gas-oil contract	GOC
Gas-oil ratio	GOR
Gasoline	gaso
Gasoline plant	GP
Gas pay	GP
Gas purchase contract	GPC
Gas Rock	G Rk
Gas sales contract	GSC
Gas show	GS
Gas too small to measure	GTSTM
Gas to surface (time)	GTS
Gastropod	gast
Gas Unit	GU
Gas volume	GV
Gas volume not measured	GVNM
Gas-water contact	GWC
Gas well	GW

Gas-well gas	GWG
Gas well shut-in	GSI
Gathering line	G/L
General	genl
General Land Office (Texas)	GLO
Generator	gen
Geological	Geol
Geologist	Geol
Geology	Geol
Geophysical	Geop
Geophysic	Geop
Georgetown	Geo
Gibson	Gib
Gilcrease	Gilc
Gilsonite	gil
Glass	gls
Glassy	gl
Glauconite	glau
Glauconitic	glau
Glen Dean lime	GD
Glen Rose	GR
Glenwood	Glen
Globigerina	Glob
Glorieta	Glor
Glycol	glyc
Gneiss	gns
Going in hole	GIH
Golconda lime	Gol
Good	gd
Good fluorescence	GFLU
Good odor & taste	gd o&t
Good show of gas	GSG
Good show of oil	GSO
Goodland	Gdld, Good L
Goodwin	Gdwn
Goose egg	g egg
Gorham	Gor
Gouldbusk	Gouldb
Government	govt
Governor	gov

Grade	gr
Grading	grdg
Grading location	grd loc
Gradual	grad
Gradually	grad
Grain	gr
Grained (as in fine-grained)	gnd
Grains per gallon	GPG
Gram	g
Gram	gm
Gram-calorie	g-cal
Gram-calorie	gm-cal
Gram molecular weight	g mole
Graneros	Granos
Granite	gran
Granite Wash	Gran W
Grant (of land)	grt
Granular	gran
Granular	grnlr
Graptolite	grap
Grating	grtg
Gravel	gvl
Gravel packed	GVLPK
Gravitometer	grvt
Gravity	grav
Gravity	GTY
Gravity, °API	gr API
Gravity meter	GM
Gray	gry
Grayburg	Grayb
Gray sand	Gr Sd
Grayson	Gray
Graywacke	gywk
Grease	gr
Greasy	gsy
Green	grn
Greenhorn	GH
Green River	Grn Riv
Green shale	Grn sh
Gritty	grty

Grooved	grv
Grooved ends	GE
Gross	grs
Gross acre feet	GAF
Gross royalty	gr roy
Gross weight	gr wt
Ground	gr
Ground	grd
Ground joint	GJ
Ground level	GL
Ground measurement (elevation)	GM
Guard log	GRDL
Gull River	G. Riv
Gummy	gmy
Gun barrel	GB
Gun Perforate	G/P
Gunsite	Guns
Gypsiferous	gypy
Gypsum	gyp
Gypsum Springs	Gyp Sprgs
Gyroidina	Gyr
Gyroidina scal	Gyr sc

Hackberry	Hackb
Hackly	hky
Hand-control valve	HCV
Hand hole	HH
Handle	hdl
Hanger	hgr
Haragan	Hara
Harbor	hbr
Hard	hd
Hardinsburg sand (local)	Hberg
Hard lime	hd li
Hardness	hdns

Hard sand	hd sd
Hardware	hdwe
Haskell	Hask
Haynesville	Haynes
Hazardous	haz
Head	hd
Header	hdr
Headquarters	HQ
Heater	htr
Heat exchanger	HEX
Heat exchanger	HX
Heat treated	ht
Heater treater	ht
Heating oil	HO
Heavy	hvy
Heavily	hvly
Heavily gas-cut mud	HGCM
Heavily gas-cut water	HGCW
Heavily oil and gas-cut mud	HO & GCM
Heavily oil-cut mud	HOCM
Heavily oil-cut water	HOCW
Heavily water-cut mud	HWCM
Heavy cycle oil	HCO
Heavy duty	HD
Heavy fuel oil	HFO
Heavy oil	HO
Heavy steel drum	HSD
Heebner	Heeb
Height	hgt
Heirs	hrs
Held by production	HBP
Hematite	hem
Herington	Her
Hermosa	Herm
Hertz (new name for electrical cycles per second)	Hz
Heterostegina	het
Hexagon(al)	hex
Hexane	hex
Hickory	Hick

High detergent	HD
High gas-oil ratio	HGOR
High-level shut-down	HLSD
Highly	hily
High pressure	HP
High-pressure gas	HPG
High-resolution dipmeter	HRD
High temperature	ht
High-temperature shut-down	HTSD
High tension	ht
High viscosity	HV
High viscosity index	HVI
Highway	hwy
Hilliard	Hill
Hockelyensis	hock
Hogshooter	Hog
Holes per foot	HPF
Hole full of oil	HFO
Hole full of salt water	HFSW
Hole full of sulphur water	HF Sul W
Hole full of water	HFW
Hollandberg	Holl
Home Creek	Home Cr
Hookwall packer	HWP
Hoover	Hov
Hopper	hop
Horizontal	horiz
Horsepower	HP
Horsepower hour	hp hr
Hospah	Hosp
Hot-rolled steel	HRS
Hour	hr
Hours	hr
Housebrand (regular grade of gasoline)	HB
Hoxbar	Hox
Humblei	Humb
Humphreys	Hump
Hundred weight	cwt
Hunton	Hun
Hydraulic	HYD

Hydraulic horsepower	HHP
Hydraulic pump	HP
Hydril	HD
Hydril thread	HYD
Hydril Type A joint	HYDA
Hydril Type CA joint	HYDCA
Hydril Type CS joint	HYDCS
Hydrocarbon	HC
Hydrofining	HFG
Hydrostatic head	HH
Hydrostatic pressure	HP
Hydrotreater	hydtr
Hygiene	Hyg

I

Identification sign	I.D. Sign
Idiomorpha	Idio
Igneous	Ign
Imbedded	imbd
Immediate(ly)	immed
Impervious	imperv
Imperial	Imp
Imperial gallon	Imp gal
Impression block	IB
Inbedded	inbdd
Incandescent	incd
Inch(es)	in
Inches mercury	in. Hg
Inches per second	in/sec
Inch-pound	in-lb
Incinerator	incin
Include	incl
Included	incl
Including	incl
Inclusions	incls
Incorporated	Inc

Increase (d) (ing)	incr
Indicate	indic
Indicated horsepower	IHP
Indicated horsepower hour	IHPHR
Indicates	indic
Indication	indic
Indistinct	indst
Individual	indiv
Induction	ind
Indurated	indr
Inflammable liquid	Inf L
Inflammable solid	Inf S
Information	info
Inhibitor	inhib
Initial	init
Initial air blow	IAB
Initial boiling point	IBP
Initial bottom-hole pressure	IBHP
Initial bottom-hole pressure flowing	IBHPF
Initial bottom-hole pressure shut-in	IBHPSI
Initial flowing pressure	IFP
Initial hydrostatic pressure	IHP
Initial mud weight	IMW
Initial potential	IP
Initial potential flowed	IPF
Initial production flowed(ing)	IPF
Initial pressure	IP
Initial production	IP
Initial production gas lift	IPG
Initial production on intermitter	IPI
Initial production plunger lift	IPL
Initial production pumping	IPP
Initial production swabbing	IPS
Initial shut-in pressure (DST)	ISIP
Initial vapor pressure	IVP
Injected	inj
Injection	inj
Injection gas-oil ratio	IGOR
Injection pressure	Inj Pr
Injection rate	IR

Injection well	IW
Inland	inl
Inlet	inl
Inoceramus (paleo)	Inoc
Input/output	I/O
Inside diameter	ID
Inside screw (valve)	IS
Inspect	insp
Inspected	insp
Inspecting	insp
Inspection	insp
Installation(s)	instl
Install (ed) (ing)	inst
Installed pumping equipment	INPE
Installing pumping equipment	INPE
Install(ing) pumping equipment	IPE
Instantaneous	inst
Instantaneous shut-in pressure (frac)	ISIP
Institute	inst
Instrument	instr
Instrumentation	instr
Insulate	ins
Insulate	insul
Insulation	ins
Insurance	ins
Integrator	intgr
Intention to drill	ITD
Interbedded	inbd
Interbedded	interbd
Intercooler	incolr
Inter-crystalline	inter-xln
Interest	int
Integral joint	IJ
Intergranular	ingr
Intergranular	inter-gran
Interior	int
Interlaminted	inlam
Interlaminated	inter-lam
Internal	int
Internal flush	IF

Internal Revenue Service	IRS
Internal upset ends	IUE
Intersect	ints
Interstitial	intl
Interval	int
Interval	intv
Intrusion	intr
Inventory	inven
Invert	inv
Invertebrate	invrtb
Inverted	inv
Invoice	inv
Ireton	Ire
Iridescent	irid
Iron body brass (bronze) mounted (valve)	IBBM
Iron body brass core (valve)	IBBC
Iron body (valve)	IB
Iron case	IC
Iron pipe size	IPS
Iron pipe thread	IPT
Ironstone	Fe-st
Ironstone	irst
Irregular	irreg
Isometric	isom
Isopropyl alcohol	IPA
Isothermal	isoth
Iverson	Ives

J

Jacket	jac
Jackson	Jack
Jackson	Jxn
Jackson sand	Jax
Jammed	jmd
Jasper(oid)	Jasp

Jefferson	Jeff
Jelly-like colloidal suspension	gel
Jet perforated	JP
Jet perforations per foot	JP/ft
Jet propulsion fuel	JP fuel
Jet shots per foot	JSPF
Jobber	jbr
Job complete	JC
Joint interest non-operated (property)	JINO
Joint operating agreement	JOA
Joint operating provisions	JOP
Joint operation	J/O
Joint(s)	jt(s)
Joint venture	JV
Jordan	Jdn
Judith River	Jud R
Juction	jct
Junction box	JB
Junk basket	JB
Junk(ed)	jnk
Junked and abandoned	J&A
Jurassic	Jur
Jurisdiction	juris

K

Kaibab	Kai
Kansas City	KC
Kaolin	kao
Kayenta	Kay
Keener	Ke
Kelly bushing	KB
Kelly bushing measurement	KBM
Kelly drive bushing	KDB
Kelly drive bushing elevation	KDBE
Kelly drill bushing to landing flange	KDB–LDG FLG

Kelly drill bushing to mean low water	KDB–MLW
Kelly drill bushing to platform	KDB–Plat
Kelvin (temperature scale)	K
Keokuk-Burlington	Keo-Bur
Kerosine	kero
Ketone	ket
Keystone	Key
Kiamichi	Kia
Kibbey	Kib
Kicked off	KO
Kickoff point	KOP
Killed	kld
Kill(ed) well	KW
Kiln dried	KD
Kilocalorie	kcal
Kilocycle	kc
Kilogram	kg
Kilogram calorie	kg-cal
Kilogram meter	kg-m
Kilohertz (See Hz-Hertz)	KHz
Kilometer	km
Kilovar-hour	kvar hr
Kilovar; reactive kilovolt-ampere	kvar
Kilovolt	kv
Kilovolt-ampere	kva
Kilovolt-ampere-hour	kvah
Kilowatt	kw
Kilowatt hour	kwh
Kilowatt-hourmeter	kwhm
Kincaid lime	Kin, KD
Kinderhook	Khk
Kinematic	Kin
Kinematic viscosity	KV
Kirtland	Kirt
KMA sand	KMA
Knock out	KO
Kootenai	Koot
Krider	Kri

L

Labor	Lab
Laboratory	Lab
Ladder	lad
Laid down	LD
Laid down cost	LDC
Laid (laying) down drill collars	LDDCs
Laid (laying) down drill pipe	LDDP
Lakota	Lak
Laminated	lam
Laminations	lam
La Motte	La Mte
Land(s)	ld(s)
Landulina	Land
Lansing	Lans
Lap joint	LJ
Lapweld	LW
Laramie	Lar
Large	lg
Large	lrg
Large Discorbis	Lg Disc
Latitude	lat
Lauders	Laud
Layton	Layt
Leached	lchd
Leadville	Leadv
League	Lge
Leak	lk
Lease	lse
Lease automatic custody transfer	LACT
Lease crude	LC
Lease use (gas)	LU
Leavenworth	Lvnwth
Le Comptom	Le C
Left hand	LH
Left in hole	LIH
Legal subdivision (Canada)	LSD

Length	lg
Lennep	Len
Lense	lns
Lenticular	len
Less-than-carload lot	LCL
Less than truck load	LTL
Letter	ltr
Level	lvl
Level alarm	LA
Level controller	LC
Level control valve	LCV
Level glass	lg
Level indicator	LI
Level indicator controller	LIC
Level recorder	LR
Level recorder controller	LRC
License	lic
Liebuscella	Lieb
Light	lt
Light barrel	LB
Light brown oil stain	LBOS
Lightening avvester	LA
Light fuel oil	LFO
Lighting	ltg
Light iron barrel	LIB
Light iron grease barrel	LIGB
Light steel drum	LSD
Lignite	lig
Lignitic	lig
Lime	li
Lime	lm
Limestone	li
Limestone	ls
Limit	lim
Limited	ltd
Limonite	lim
Limy	lmy
Limy shale	Lmy sh
Linear	lin
Linear foot	lin ft

Line, as in E/L (East line)	/L
Line pipe	L.P.
Liner	lnr
Liner	lin
Linguloid	lngl
Liquefaction	liqftn
Liquefied natural gas	LNG
Liquefied petroleum gas	LP-Gas
Liquid	liq
Liquid level controller	LLC
Liquid level gauge	LLG
Liquid volume	LV
Liter	l
Lithographic	litho
Little	ltl
Load	ld
Load acid	LA
Load oil	LO
Load water	LW
Local purchase order	LPO
Located	loc
Location	loc
Location abandoned	loc abnd
Location graded	loc gr
Lock	lk
Locker	lkr
Lodge pole	LP
Log mean temperature difference	LMTD
Log total depth	LTD
Long	lg
Long coupling	LC
Long handle/round point	LH/RP
Longitude(inal)	long
Long radius	LR
Long threads & coupling	LT&C
Loop	lp
Lost circulation	LC
Lost circulation material	LCM
Lovell	Lov
Lovington	Lov

Low pressure	LP
Low-pressure separator	LP sep
Low-temperature extraction unit	LTX unit
Low-temperature separation unit	LTS unit
Low-temperature shut down	LTSD
Low viscosity index	LVI
Lower	low
Lower	lwr
Lower Albany	L/Alb
Lower anhydrite stringer	LAS
Lower, as L/Gallup	L/
Lower Cretaceous	L/Cret
Lower explosive limit	LEL
Lower Glen Dean	LGD
Lower Menard	LMn
Lower Tuscaloosa	L/Tus
Lube oil	LO
Lubricant	lub
Lubricate (d) (ing) (ion)	lub
Lueders	Lued
Lug cover type (5-gallon can)	LC
Lug cover with pour spout	LCP
Lumber	lbr
Lumpy	lmpy
Lustre	lstr

M

Macaroni tubing	MT
Machine	mach
Mackhank	Mack
Madison	Mad
Magnetic	mag
Magnetometer	mag
Maintenance	maint
Major	maj
Majority	maj

Male and female (joint)	M&F
Male pipe thread	MPT
Male to female angle	MFA
Malleable	mall
Malleable iron	MI
Management	m'gmt
Manager	mgr
Manhole	MH
Manifold	man
Manifold	MF
Manitoban	Manit
Manning	Mann
Manual	man
Manually operated	man op
Manufactured	mfd
Manufacturing	mfg
Maquoketa	Maq
Marble Falls	Mbl Fls
Marchand	March
Marginal	marg
Marginulina	Marg
Marginulina coco	Marg coco
Marginulina flat	Marg fl
Marginulina round	Marg rd
Marginulina texana	Marg tex
Marine	mar
Marine gasoline	margas
Marine rig	MR
Marine terminal	M/T
Marine Tuscaloosa	M'Tus
Marine wholesale distributors	MWD
Market(ing)	mkt
Market demand factor	MDF
Markham	Mark
Marlstone	mrlst
Marly	mly
Marmaton	Marm
Maroon	mar
Massilina pratti	Mass pr
Massive	mass

Massive Anhydrite	MA
Master	mstr
Material	matl
Material	mtl
Mathematics	math
Matrix	mtx
Matter	mat
Maximum	max
Maximum & final pressure	M&FP
Maximum daily delivery obligation	MDDO
Maximum efficient rate	MER
Maximum flowing pressure	MFP
Maximum operating pressure	MOP
Maximum pressure	MP
Maximum surface pressure	MSP
Maximum top pressure	MTP
Maximum tubing pressure	MTP
Maximum working pressure	MWP
Maywood	May
McClosky lime	McC
McCullough	McCul
McElroy	McEl
McKee	McK
McLish	McL
McMillan	McMill
Meakin	Meak
Mean effective pressure	MEP
Mean low water to platform	MLW–PLAT
Mean temperature difference	MTD
Measure (ed) (ment)	meas
Measured depth	MD
Measured total depth	MTD
Measuring and regulating station	M & R Sta.
Mechanic (al) (ism)	mech
Mechanical down time	Mech DT
Mechanism	mchsm
Median	med
Medicine Bow	Med B
Medina	Med
Medium	med

Medium amber cut	MAC
Medium fuel oil	med FO
Medium-grained	m-gr
Medium-grained	med gr
Medrano	Medr
Meeteetse	Meet
Megacycle	mc
Megahertz (megacycles per second)	MHz
Melting point	MP
Member (geologic)	mbr
Memorandum	memo
Menard lime	Men
Menefee	Mene
Meramec	Mer
Mercaptan	mercap
Mercaptan	RSH
Merchandise	mdse
Mercury	merc
Meridian	merid
Mesaverde	mvde
Mesozoic	Meso
Metal petal basket	MPB
Metamorphic	meta
Meter	m
Meter	mtr
Meter run	MR
Methane	meth
Methane-rich gas	MRG
Methanol	methol
Methyl chloride	meth-cl
Methyl ethyl ketone	MEK
Methyl isobutyl ketone	MIK
Methylene blue	MB, meth-bl
Metric	metr
Mezzanine	mezz
Mica	mic
Micaceous	mic
Microcrystalline	mcr-x
Microcrystalline	micro-xin
Microfarad	mfd

Microfossil(iferous)	micfos
Microwave	MW
Middle	M/
Middle	mdl
Middle	mdl
Midway	Mid
Midway	Mwy
Mile(s)	MI
Miles per hour	MPH
Military	mil
Milky	mky
Mill wrapped plain end	MWPE
Milled	mld
Milled other end	MOE
Milliampere	ma
Millidarcies	md
Milligram	mg
Millihenry	mh
Milliliter	ml
Millimeters of mercury	mm HG
Milliliters tetrethyl lead per gallon	m! TEL
Millimeter	mm
Milling	millg
Milling	mlg
Milliolitic	mill
Million	mil
Million British thermal units	MMBTU
Million cubic feet	MMCF
Million cubic feet/day	MMCFD
Million electron volts	mev
Million standard cubic feet per day	MMSCFD
Millivolt	mv
Mineral	mnrl
Minerals	min
Mineral interest	MI
Minimum	min
Minimum pressure	min P
Minnekahta	Mkta
Minnelusa	Minl
Minute(s)	min

Miocene	Mio
Miscellaneous	misc
Misener	Mise
Mission Canyon	Miss Cany
Mississippian	Miss
Mixed	mxd
Mixer	mix
Mobile	mob
Model	mod
Moderate(ly)	mod
Modification	mod
Modular	modu
Moenkopi	Moen
Moisture, impurities, and unsaponi-fiables (grease testing)	MIU
Molas	mol
Mole	mol
Molecular weight	mol wt
Mollusca	mol
Monitor	mon
Montoya	Mont
Moddy's Branch	MB
Moore County Lime	MC Ls
Mooringsport	Moor
More or less	m/l
Morrison	Morr
Morrow	Mor
Mortgage	mtge
Mosby	Mos
Motor	mot
Motor generator	MG
Motor medium	MM
Motor octane number	MON
Motor oil	MO
Motor oil units	MOU
Motor severe	MS
Motor vehicle	M/V
Motor vehicle fuel tax	MVFT
Motor vessel	M/V
Mottled	mott

Mount Selman	Mt. Selm
Mounted	mtd
Mounting	mtg
Moving	mov
Moving in (equipment)	MI
Moving in and rigging up	MIRU
Moving in cable tools	MICT
Moving in completion unit	MICU
Moving (moved) in double drum unit	MIDDU
Moving in materials	MIM
Moving in pulling unit	MIPU
Moving in rig	MIR
Moving in rotary tools	MIRT
Moving in service rig	MISR
Moving in standard tools	MIST
Moving in tools	MIT
Moving out	MO
Moving out cable tools	MOCT
Moving out completion unit	MOCU
Moving out rig	MOR
Moving out rotary tools	MORT
Mowry	Mow
Mud acid	MA
Mud acid wash	MAW
Mud cleanout agent	MCA
Mud cut	MC
Mud filtrate	MF
Mud logging unit	MLU
Mud logger	ML
Mud to Surface	MTS
Mud weight	md wt
Mud Weight	mud wt
Mud-cut acid	MCA
Mud-cut oil	MCO
Mud-cut salt water	MCSW
Mud-cut water	MCW
Muddy	Mdy
Muddy water	MW
Mudstone	mudst
Multi-grade	MG

Multiple service acid	MSA
Multipurpose	MP
Multipurpose grease, lithium base	MPGR-Lith
Multipurpose grease, soap base	MPGR-soap
Muscovite	musc

N

Nacatoch	Nac
Nacreous	nac
Nameplate	NP
Naphtha	nap
National	Nat'l
National coarse thread	NC
National Electric Code	NEC
National Fine (thread)	NF
National pipe thread	NPT
National pipe thread, female	NPTF
National pipe thread, male	NPTM
Natural	nat
Natural flow	NF
Natural gas	NG
Natural gas liquids	NGL
Nautical mile	NMI
Navajo	Nav
Navarro	Navr
Negative	neg
Negligible	neg
Neoprene	npne
Net effective pay	NEP
Net tons	NT
Neutral	neut
Neutralization	neut
Neutralization Number	Neut. No.
New Albany shale	New Alb
New bit	NB
Newburg	Nbg

Newcastle	Newc
New field discovery	NFD
New field wildcat	NFW
New oil	NO
New pool discovery	NPD
New pool exempt (nonprorated)	NPX
New pool wildcat	NPW
New total depth	NTD
Niagara	Nig
Nickle plated	NP
Ninnescah	Nine
Niobrara	Niob
Nipple	nip
Nipple (d) (ing) up blowout preventers	NUBOPs
Nipple (d) (ing) down blowout preventers	NDBOPs
Nippling up	NU
Nitrogen blanket	NB
Nitroglycerine	nitro
No appreciable gas	NAG
Noble-Olson	NO
No change	NC
No core	NC
Nodosaria blanpiedi .	Nod Blan
Nodosaria Mexicana	Nod mex
Nodular	nod
Nodule	nod
No fluorescence	NF
No fluid	NF
No fuel	NF
No gas to surface	NGTS
No gauge	NG
No increase	No Inc
Nominal	nom
Nominal pipe size	NPS
Non-contiguous tract	NCT
Non-destructive testing	NDT
Non-detergent	ND
Nonemulsifying agent	NE
Non-emulsion acid	NEA
Nonflammable compressed gas	nonf G

Nonionella	Non
Nonionella Cockfieldensis	N. Cock.
Non-leaded gas	NL Gas
Non-operated joint ventures	NOJV
Non-operating property	NOP
Non-porous	NP
Non-returnable	NR
Non-returnable steel barrel	NRSB
Non-returnable steel drum	NRSD
Non-rising stem (valve)	NRS
Non-standard	nstd
Non-standard service station	N/S S/S
Non-upset	NU
Non-upset ends	NUE
Noodle Creek	Ndl Cr
No order required	NOR
No paint on seams	npos
No production	NP
No recovery	no rec
No recovery	NR
No report	NR
No returns	NR
Normal	nor
Normally closed	NC
Normally open	NO
Northeast	NE
Northeast corner	NEC
Northeast line	NEL
Northeast quarter	NE/4
Northerly	N'ly
Northern Alberta land registration district	NALRD
North half	N/2
North line	NL
North offset	N/O
North quarter	N/4
Northwest	NW
Northwest corner	NW/C
Northwest line	NWL
Northwest quarter	NW/4

Northwest Territories	NWT
No show	NS
No show gas	NSG
No show oil	NSO
No show oil & gas	NSO&G
Not drilling	ND
Not applicable	NA
Notary public	NP
Not available	NA
No test	N/tst
No time	NT
Not in contract	NIC
Not prorated	NP
Not pumping	NP
Not reported	NR
Not to Scale	NTS
Not yet available	NYA
No visible porosity	NVP
No water	NW
Nozzle	noz
Nugget	Nug
Number	NO
Numerous	num

O

Oakville	Oakv
Object	obj
Obsolete	obsol
Occasional(ly)	occ
Ocean bottom suspension	OBS
Octagon	oct
Octagonal	oct
Octane	oct
Octane number requirement	ONR
Octane number requirement increase	ONRI
O'Dell	Odel

Odor	od
Odor, stain & fluorescence	O, S & F
Odor, taste & stain	O, T & S
Odor, taste, stain & fluorescence	O, T, S & F
Off bottom	OB
Office	off
Official	off
Official potential test	OPT
Off-shore	off-sh
O'Hara	O'H
Ohm-centimeter	ohm-cm
Ohm-meter	ohm-m
Oil	O
Oil and gas	O&G
Oil and Gas Journal	OGJ
Oil and gas lease	O&GL
Oil and gas-cut mud	O&GCM
Oil and gas-cut salt water	O&GCSW
Oil & gas-cut sulphur water	O&GC SULW
Oil and gas-cut water	O&GCW
Oil and salt water	O&SW
Oil & sulphur water-cut mud	O&SWCM
Oil and water	O&W
Oil base mud	OBM
Oil change	O/C
Oil circuit breaker	OCB
Oil Creek	Oil Cr
Oil cut	OC
Oil emulsion	OE
Oil emulsion mud	OEM
Oil fluorescence	OFLU
Oil in hole	OIH
Oil in place	OIP
Oil odor	OO
Oil pay	OP
Oil payment interest	OPI
Oil sand	O sd
Oil show	OS
Oil soluble acid	OSA

Oil stain	OSTN
Oil string flange	OSF
Oil to surface	OTS
Oil unit	OU
Oil well flowing	OWF
Oil well shut-in	OSI
Oil-cut mud	OCM
Oil-cut salt water	OCSW
Oil-cut water	OCW
Oil-powered total energy	OTE
Oil-water contact	OWC
Oil-well gas	OWG
Old abandoned well	OAW
Old plug-back depth	OPBD
Old total depth	OTD
Old well drilled deeper	OWDD
Old well plugged back	OWPB
Old well worked over	OWWO
Olefin	ole
Oligocene	Olig
On center	OC
One thousand foot-pounds	kip-ft
One thousand pounds	kip
Oolicastic	ooc
Oolimoldic	oom
Oolitic	ool
Open (ing) (ed)	opn
Open cup	OC
Open end	OE
Open flow	OF
Open flow potential	OFP
Open hearth	OH
Open hole	OH
Open tubing	Ot
Operate	oper
Operations	oper
Operations commenced	OC
Operator	oper
Operculinoides	Operc.
Opposite	opp
Optimum bit weight and rotary speed	OBW & RS

Option to farmout	optn to F/O
Ordovician	Ord
Oread	Or
Orifice	orf
Orifice flange one end	OFOE
Organic	org
Organization	org
Original	orig
Originally	orig
Original stock tank oil in place	OSTOIP
Oriskany	Orisk
Orthoclase	orth
Osage	Os
Osborne	Osb
Ostracod	Ost
Oswego	Osw
Other end beveled	OEB
Ounce	oz
Ouray	Our
Out of service	O/S
Out of stock	O/S
Outboard motor oil	OBMO
Outer continental shelf	OCS
Outlet	otl
Outpost	OP
Outside diameter	OD
Outside screw and yoke (valve)	OS&Y
Over and short (report)	O/S
Over produced	OP
Overall	OA
Overall height	OAH
Overall length	OAL
Overhead	OH
Overhead	ovhd
Overriding royalty	ORR
Overriding royalty interest	ORRI
Overshot	OS
Overtime	OT
Oxidation	ox
Oxidized	ox
Oxygen	oxy

P

Package	pkg
Packaged	pkgd
Packed	pkd
Packer	pkr
Packer set at	PSA
Packing	pkg
Paddock	Padd
Pahasapa	Paha
Paid	PD
Paint Creek	PC
Pair	pr
Paleontology	paleo
Paleozoic	Paleo
Palo Pinto	Palo P
Paluxy	Pal
Paluxy	Pxy
Panel	pnl
Panhandle Lime	Pan L
Paradox	Para
Paraffins, olefins, naphthenes, aromatics	PONA
Parish	Ph
Park City	Park C
part	pt
Partings	prtgs
Partly	prtly
Partly	pt
Parts per million	ppm
Patent (ed)	pat
Pattern	patn
Pavement	pvmnt
Paving	pav
Pawhuska	Paw
Payment	PP
Payment	payt
Pearly	prly
Pebble	pbl

Pebbles	pebs
Pebbly	pbly
Pecan Gap	PG
Pecan Gap Chalk	PGC
Pelecypod	plcy
Pelletal	pell
Palletoidal	pell
Penalize, penalized, penalizing, penalty	penal
Penetration test	pen
Penetration	pen
Penetration asphalt cement	Pen A.C.
Penetration index	PI
Pennsylvanian	Penn
Pensky Martins	PM
Per acre bonus	PAB
Per acre rental	PAR
Per day	PD
Percent	pct
Percolation	perco
Perforate (d) (ing) (or)	perf
Perforated casing	perf csg
Performance Number (Av gas)	PN
Period	prd
Permanent	perm
Permanently shut down	PSD
Permeability	perm
Permeable	perm
Permian	Perm
Permit	prmt
Perpendicular	perp
Personal and confidential	P&C
Personnel	pers
Petrochemical	petrochem
Petroleum	pet
Petroleum & natural gas	P & NG
Petroliferous	petrf
Pettet	Pet
Pettit	Pett
Pettus sd	Pet sd
Phase	ph

Phosphoria	Phos
Phrrhotite	po
Picked up	PU
Pictured Cliff	Pic Cl
Piece	pc
Pieces	pcs
Pilot	plt
Pin Oak	P.O.
Pine Island	PI
Pink	pk
Pinpoint	PP
Pinpoint	pinpt
Pinpoint porosity	PPP
Pint	pt
Pipe, buttweld	PBW
Pipe, electric weld	PEW
Pipe, lapweld	PLW
Pipe, seamless	PSM
Pipe, spiral weld	PSW
Pipe to soil potential	PTS pot.
Pipeline	PL
Pipeline oil	PLO
Pipeline terminal	PLT
Pisolites	piso
Pisolitic	piso
Pitted	pit
Plagioclase	plag
Plain end	PE
Plain end beveled	PEB
Plain large end	PLE
Plain one end	POE
Plain small end	PSE
Plan	pln
Plant	plt
Plant fossils	pl fos
Plant volume reduction	PVR
Planulina harangensis	Plan. harangensis
Planulina palmarie	Plan. palm.
Plastic	plas

Plastic viscosity	P V
Platform	platf
Platformer	platfr
Platy	plty
Please note and return	PNR
Pleistocene	Pleist
Pliocene	Plio
Plug on bottom	POB
Plugged	plgd
Plugged & abandoned	P&A
Plugged Back	PB
Plugged back depth	PBD
Plugged back total depth	PBTD
Plunger	plngr
Pneumatic	pneu
Podbielniak	Pod.
Point	pt
Point Lookout	Pt Lkt
Poison	pois
Poker chipped	PC
Polish(ed)	pol
Polished rod	PR
Polyethylene	polyel
Polymerization	poly
Polymerized	poly
Polymerized gasoline	polygas
Polypropylene	polypl
Polyvinyl chloride	poly cl
Polyvinyl chloride	PVC
Pontotoc	Pont
Pooling agreement	PA
Porcelaneous	porc
Porcion	porc
Pore volume	P.V.
Porosity	por
Porosity and permeability	P&P
Porous	por
Porous and permeable	P&P
Portable	port
Porter Creek	PC

Position	pos
Positive	pos
Positive crankcase ventilation	PCV
Possible(ly)	poss
Post Laramie	P Lar
Post Oak	P.O.
Potential	pot
Potential difference	pot dif
Potential test	PT
Potential test to follow	PTTF
Pound	lb
Pound per foot	lb/ft
Pound-inch	lb-in
Pounds per gallon	ppg
Pounds per square foot	lb/sq ft
Pounds per square foot	psf
Pounds per square inch	psi
Pounds per square inch absolute	psia
Pounds per square inch gauge	psig
Pour point (ASTM Method)	pour ASTM
Power	PWR
Power factor	PF
Power factor meter	PFM
Pre-Cambrian	Pre Camb
Precast	prcst
Precipitate	ppt
Precipitation Number	pptn No
Predominant	predom
Prefabricated	prefab
Preferred	pfd
Preheater	prehtr
Preliminary	prelim
Premium	prem
Prepaid	ppd
Preparation	Prep
Prepare	Prep
Preparing	Prep
Preparing to take potential test	PRPT
Pressed distillate	PD
Pressure	press

Pressure alarm	PA
Pressure control valve	PCV
Pressure differential controller	PDC
Pressure differential indicator	PDI
Pressure differential indicator controller	PDIC
Pressure differential recorder	PDR
Pressure differential recorder controller	PDRC
Pressure indicator	PI
Pressure indicator controller	PIC
Pressure recorder	PR
Pressure recorder control	PRC
Pressure switch	PS
Pressume-volume-temperature	PVT
Prestressed	prest
Prevent	prev
Preventive	prev
Primary	pri
Primary Reference Fuel	PRF
Principal	prin
Principal lessee(s)	prncpl lss
Prism(atic)	pris
Private branch exchange	PBX
Privilege	priv
Probable(ly)	prob
Process	proc
Produce (d) (ing) (tion)	prod
Producing gas well	PGW
Producing oil & gas well	POGW
Producing oil well	POW
Producing oil well flowing	POWF
Producing oil well pumping	POWP
Product(s)	prod
Production	PP
Production department exploratory test	PDET
Production payment interest	PPI
Productivity index	PI
Profit & loss	P & L
Profit sharing interest	PSI
Progress	prog
Project (ed) (ion)	proj

Property line	PL
Proportional	prop
Propose(d)	prop
Proposed bottom hole location	PBHL
Proposed depth	PD
Prorated	pro
Protection	prot
Proterozoic	Protero
Provincial	Prov
Pryoclastic	pyrclas
Pseudo	pdso
Pseudo	ps
Public relations	PR
Public School Land	PSL
Pull(ed) rods & tubing	PR&T
Pulled	pld
Pulled big pipe	PBP
Pulled out	PO
Pulled out of hole	POH
Pulled pipe	PP
Pulled up	PU
Pulling	plg
Pulling tubing	PTG
Pulling tubing and rods	PTR
Pump, pumped, pumping	pmp (d) (g)
Pump and flow	P&F
Pump jack	PJ
Pump job	PJ
Pump on beam	POB
Pump-in pressure	PIP
Pumping equipment	PE
Pumping for test	PFT
Pumping load oil	PLO
Pumping unit	PU
Pumps off	PO
Purchase order	PO
Purple	purp
Putting on pump	POP
Pyrite, pyritic	pyr
Pyrobitumen	pyrbit
Pyrolysis	pyls

Q

Quadrangle	quad
Quadrant	quad
Quadruple	quad
Quality	qual
Quantity	qty
Quantity	quan
Quantity discount allowance	QDA
Quarry	qry
Quart(s)	qt
Quarter	qtr
Quartz	qtz
Quartzite	qtz
Quartzitic	qtz
Quartzose	qtzose
Queen City	Q. City
Queen Sand	Q. Sd
Quench	qnch
Questionable	quest
Quick ram change	QRC
Quintuplicate	quint

R

Radial	rad
Radian	RAD
Radiation	radtn
Radioactive	RA
Radiological	RAD
Radius	RAD
Railing	rlg
Railroad	RR
Railroad Commission (Texas)	RRC

Railway	Ry
Rainbow show of oil	RBSO
Raised faced	RF
Raised face flanged end	RFFE
Raised face weld neck	RFWN
Ramsbottom Carbon Residue	RCR
Ran in hole	RIH
Ran (running) rods and tubing	RR&T
Random lengths	RL
Range	R
Range	Rge
Ranger	Rang
Rankine (temp. scale)	R
Rapid curing	RC
Rat hole	RH
Rat hole mud	RHM
Rate of penetration	ROP
Rate of return	ROR
Rate too low to measure	RTLTM
Rating	rtg
Reacidize	reacd
Reacidized	reacd
Reacidizing	reacd
Reactor	reac
Reagan	Reag
Ream	RM
Reamed	rmd
Reaming	rmg
Reboiler	reblr
Received	recd
Receptacle	recp
Reciprocate(ing)	recip
Recirculate	recirc
Recommend	rec
Recomplete (d) (ion)	recomp
Recondition(ed)	recond
Recorder	rec
Recording	rec
Recover	rec
Recovered	rec

Recovering	rec
Recovery	rec
Rectangle	rect
Rectangular	rect
Rectifier	rect
Recycle	recy
Red Beds	Rd Bds
Red cave	RC
Red Fork	Rd Fk
Red indicating lamp	RIL
Red Oak	R.O.
Red Peak	Rd Pk
Red River	RR
Redrilled	redrld
Reducer	red
Reducing	red
Reducing balance	red bal
Reference	ref
Refine (d) (r) (ry)	ref
Refining	refg
Reflection	refl
Reflux	refl
Reformate	reform
Reformer	reform
Reforming	reform
Refraction	refr
Refractory	refr
Refrigerant	referg
Refrigeration	refer
Refrigerator	refgr
Regenerator	regen
Regular acid	R/A
Reid vapor pressure	RVP
Reinforce (d)ng)	reinf
Reinforced concrete	reinf conc
Reinforcing bar	rebar
Reject	rej
Rejection	rej'n
Reklaw	Rek
Register	reg

Regular	reg
Regulator	reg
Relay	rel
Relay	rly
Release(d)	rel
Release (d) (ing)	rls (d) (ing)
Released swab unit	RSU
Relief	rlf
Relocate(d)	reloc
Remains	rem
Remains	rmn
Remedial	rem
Remote control	RC
Removable	rmv
Remove (al) (able)	rem
Renault	Ren
Rental	rent
Reophax bathysiphoni	Reo bath
Repair (ed) (ing) (s)	rep
Repairman	rpmn
Reperforated	reperf
Replace(d)	rep
Replace(ment)	repl
Report	rep
Required	reqd
Requirement	reqmt
Requisition	req
Requisition	reqn
Research	res
Research and development	R&D
Research Octane Number	RON
Research Octane Number	Res. O. N.
Reservation	res
Reserve	res
Reservoir	res
Residual	resid
Residue	resid
Resinous	rsns
Resistance	res
Resistivity	res

Resistor	res
Retail pump price	RPP
Retain (ed) (er) (ing)	ret
Retainer	rtnr
Retard(ed)	ret
Retrievable bridge plug	RBP
Retrievable retainer	retr ret
Retrievable test treat squeeze (tool)	RTTS
Return	ret
Return on investment	ROI
Returnable steel drum	RSD
Returned	retd
Returned well to production	RWTP
Returning circulation oil	RCO
Reverse(d)	rev
Reverse(d)	rvs(d)
Reverse circulation	RC
Reverse circulation rig	RCR
Reversed out	RO
Reversed out	rev/O
Revise (d) (ing) (ion)	rev
Revolution(s)	rev
Revolutions per minute	rpm
Revolutions per second	rps
Rework(ed)	rwk(d)
Rheostat	rheo
Ribbon sand	Rib
Rich oil fractionator	ROF
Rierdon	Rier
Rig floor	RF
Rig on location	ROL
Rig released	RR
Rig released	rig rel
Rig skidded	RS
Rigged down	RD
Rigged down swabbing unit	RDSU
Rigged up	RU
Rigging down	RD
Rigging up	RU
Rigging up cable tools	RUCT

Rigging up machine	RUM
Rigging up pump	RUP
Rigging up rotary	RUR
Rigging up rotary tools	RURT
Rigging up service rig	RUSR
Rigging up standard tools	RUST
Rigging up tools	RUT
Right angle	RA
Right hand	RH
Right hand door	RHD
Right-of-way	R/W
Right-of-way	ROW
Ring	rg
Ring groove	RG
Ring Joint	RJ
Ring joint flanged end	RJFE
Ring tool joint	RTJ
Ring type joint	RTJ
Rising stem (valve)	RS
Rivet	riv
Road	rd
Road & location	R&L
Road & location complete	R&LC
Roads	rds
Robulus	Rob
Rock	rk
Rock bit	RB
Rock pressure	RP
Rockwell hardness number	RHN
Rocky	rky
Rodessa	Rod
Rods & tubing	R & T
Room	rm
Root mean square	RMS
Rose	ro
Rosiclare sand	Ro
Rotary	rot
Rotary bushing	RB
Rotary bushing measurement	RBM
Rotary drive bushing	RDB

Rotary drive bushing to ground	RDB-GD
Rotary table	RT
Rotary test	R test
Rotary tools	RT
Rotary unit	RU
Rotate	rot
Rotator	rot
Rough	rgh
Round	rd
Round thread	rd thd
Round trip	rd tp
Rounded	rdd
Rounded	rnd
Royalty	roy
Royalty interest	RI
Rubber	rbr
Rubber	rub
Rubber ball sand oil frac	RBSOF
Rubber ball sand water frac	RBSWF
Rubber balls	Rbls
Run of mine	ROM
Running	rng
Running casing	RC
Running electric log	REL
Running radioactive log	RALOG
Running tubing	RTG
Rupture	rupt
Rust & oxidation	R&O

S

Sabinetown	Sab
Saccharoidal	sach
Sacks	sk
Sacks	sx
Saddle	sadl
Saddle Creek	Sad Cr

Safety	saf
Saint Genevieve	St. Gen
Saint Louis Lime	St L
Saint Peter	St Ptr
Salado	Sal
Salaried	sal
Salary	sal
Saline Bayou	Sal Bay
Salinity	sal
Salt & pepper	s&p
Salt Mountain	Slt Mt
Salt wash	SW
Salt water	SW
Salt water disposal	SWD
Salt water disposal system	SWDS
Salt water disposal well	SWDW
Salt water injection	SWI
Salt water to surface	SWTS
Salty	slty
Salty sulfur water	SSUW
Salvage	salv
Sample	samp
Sample tops	S/T
San Adres	San And
San Angelo	San Ang
San Bernardino Base and Meridian	SBB&M
San Rafael	San Raf
Sanastee	Sana
Sand	sd
Sand and shale	sd & sh
Sand oil fract	sdoilfract
Sand oil fracture	SOF
Sand showing gas	Sd SG
Sand showing oil	Sd SO
Sand water fract	sdwtrfract
Sandfrac	SF
Sandfract	sdfract
Sandstone	sd
Sand-water fracture	SWF
Sandy	sdy

Sandy lime	sdy li
Sandy shale	sdy sh
Sanitary	sani
Saponification	sap
Saponification number	Sap No
Saratoga	Sara
Satanka	Stnka
Saturated	sat
Saturation	sat
Sawatch	Saw
Sawtooth	Sawth
Saybolt Furol	Say Furol
Saybolt Seconds Universal	SSU
Saybolt Universal Seconds	SUS
Scales	sc
Scattered	sctrd
Schedule	sch
Schematic	schem
Scolecodonts	scolc
Scratcher	scr
Screen	scr
Screw	scr
Screw end American National Acme thread	SE NA
Screw end American National Coarse thread	SE NC
Screw end American National Fine thread	SE NF
Screw End American National Taper Pipe Thread	SE NPT
Screwed	scrd
Screwed end	S/E
Screwed on one end	SOE
Scrubber	scrub
Seabreeze	Sea
Sealed	sld
Seamless	smls
Seating nipple	SN
Secant	sec
Second	sec
Secondary	sec
Secondary butyl alcohol	SBA

Secretary	sec
Section	sec
Section line	SL
Section-township-range	S-T-R
Securaloy	scly
Sediment (s)	sed
Sedwick	Sedw
Seismic	seis
Seismograph	seis
Selenite	sel
Self (spontaneous) potential	SP
Selma	Sel
Senora	Sen
Separator	sep
Septuplicate	sept
Sequence	seq
Serial	ser
Series	ser
Serpentine	Serp
Serratt	Serr
Service(s)	serv
Service charge	serv chg
Service station	SS
Set plug	SP
Settling	set
Seven Rivers	S Riv
Severy	Svry
Sewer	sew
Sexton	Sex
Sextuple	sxtu
Sextuplicate	sext
Shaft horsepower	shp
Shake out	SO
Shale	sh
Shaley	shly
Shallower pool (pay) test	SPT
Shannon	Shan
Shear	shr
Sheathing	shthg
Sheet	sh

Shells	shls
Shinarump	Shin
Shipment	shpt
Shipping	shpg
Short radius	SR
Short thread	ST
Short threads & coupling	ST&C
Shot open hole	SOH
Shot point	SP
Shoulder	shld
Show condensate	SC
Show gas	SG
Show gas and condensate	SG&C
Show gas and distillate	SG&D
Show gas & water	SG&W
Show of dead oil	SDO
Show of free oil	SFO
Show of gas and oil	SG&O
Show oil	SO
Show oil and gas	SO&G
Show oil and water	SO&W
Shut down	SD
Shut down awaiting orders	SDWO
Shut down for orders	SDO
Shut down for pipe line	SDPL
Shut down for repairs	SDR
Shut down for weather	SDW
Shut down overnight	SDON
Shut down to acidize	SDA
Shut down to fracture	SDF
Shut down to log	SDL
Shut down to plug & abandon	SDPA
Shut in	SI
Shut in bottom hole pressure	SIBHP
Shut in casing pressure	SICP
Shut in gas well	SIGW
Shut in oil well	SIOW
Shut in pressure	SIP
Shut in tubing pressure	SITP
Shut in well head pressure	SIWHP

Shut in–waiting on potential	SIWOP
Side door choke	SD Ck
Side opening	SO
Sideboom	SB
Siderite(ic)	sid
Sides, tops & bottoms	s, t & b
Sidetrack(ing)	ST (g)
Sidetrack (ed) (ing)	sdtkr
Sidetracked hole	STH
Sidetracked total depth	STTD
Sidewall cores	SWC
Sidewall samples	SWS
Signed	sgd
Silica	silic
Siliceous	silic
Silky	slky
Silt	slt
Siltstone	silt
Silurian	Sil
Similar	sim
Simpson	Simp
Single pole double throw	SPDT
Single pole single throw	SPST
Single random lengths	SRL
Single Shot	SS
Singles	sgls
Siphonina davisi	Siph. d.
Size	sz
Skimmer	skim
Skinner	Skn
Skull Creek	Sk Crk
Sleeve	sl
Sleeve bearing	SB
Slickensided	sks
Sliding scale royalty	S/SR
Slight(ly)	sl
Slight(ly)	sli
Slight show of gas	SSG
Slight show of oil	sli SO
Slight show of oil & gas	SSO&G

Slight show oil	SSO
Slight, weak, or poor fluorescence	SFLU
Slightly gas-cut mud	SGCM
Slightly gas-cut oil	SGCO
Slightly gas-cut water	SGCW
Slightly gas-cut water blanket	SGCWB
Slightly oil & gas-cut mud	SO&GCM
Slightly oil-cut mud	SOCM
Slightly oil-cut water	SOCW
Slightly porous	SP
Sligo	Sli
Slim hole drill pipe	SHDP
Slip on	SO
Slow set (cement)	SS
Slurry	slur
Smackover	Smk
Small	sm.
Small show	SS
Smithwick	Smithw
Smoke Volatility Index	SVI
Smooth	smth
Socket	skt
Socket weld	SW
Socket weld	sow
Sodium base grease	sod gr
Sodium carboxymethylcellulose	CMC
Soft	sft
Solenoid	slnd
Solenoid	sol
Solids	sol
Solution	soln
Solvent	solv
Somastic	som
Somastic coated	somct
Sorted (ing)	sort
Sort (ed) (ing)	srt (d) (g)
South half	S/2
South line	SL
South offset	SO
Southeast	SE

Southeast corner	SE/C
Southeast quarter	SE/4
Southwest	SW
Southwest corner	SW/c
Southwest quarter	SW/4
Spacer	spcr
Spare	sp
Sparta	Sp
Spearfish	Spf
Special	spcl
Specialty	splty
Specific gravity	sp gr
Specific heat	sp ht
Specific volume	sp. vol.
Specification	spec
Speckled	speck
Sphaerodina	Sphaer
Sphalerite	sphal
Spherules	sph
Spicule(ar)	spic
Spigot and spigot	s & s
Spindle	spdl
Spindletop	Spletp
Spiral weld	SW
Spirifers	spfr
Spiroplectammina barrowi	Spiro, b.
Splintery	splty
Split	splty
Sponge	spg
Spore	sp
Spotted	sptd
Spotty	sptty
Spraberry	Spra
Spring	spg
Springer	Sprin
Sprinkler	spkr
Sprocket	spkt
Spud (ded) (der)	spd
Square	sq
Square centimeter	sq cm

Square foot	sq ft
Square inch	sq in
Square kilometer	sq km
Square meter	sq m
Square millimeter	sq mm
Square yard(s)	sq yd
Squeeze (d) (ing)	sqz
Squeeze packer	sq pkr
Squeezed	sq
Squirrel cage	sq cg
Stabilized(er)	stab
Stain (ed) (ing)	stn (d) (g)
Stain and odor	S&O
Stainless steel	SS
Stairway	stwy
Stalnaker	Stal
Stand (s) (ing)	std (s) (g)
Stand by	stn/by
Standard	std (s) (g)
Standard cubic feet per day	SCFD
Standard cubic feet per hour	SCFH
Standard cubic feet per minute	SCFM
Standard cubic foot	SCF
Standard operational procedure	SOP
Standard temperature and pressure	STP
Stanley	Stan
Starting fluid level	SFL
State lease	SL
State potential	State pot
Static bottom-hole pressure	SBHP
Station	sta
Stationary	stat
Statistical	stat
Steady	stdy
Steam	stm
Steam cylinder oil	stm cyl oil
Steam Emulsion Number	SE No.
Steam engine oil	stm eng oil
Steam working pressure	SWP
Steel	stl
Steel line correction	SLC

Steel line measurement	SLM
Steel tape measurement	STM
Steele	Stel
Stenographer	steno
Stensvad	Stens
Sticky	stcky
Stippled	stip
Stirrup	stir
Stock	stk
Stock tank barrels	STB
Stock tank barrels per day	STB/D
Stock tank oil in place	STOIP
Stone Corral	Stn Crl
Stony Mountain	Sty Mt
Stopped	stpd
Stopper	stp
Storage	stor
Storage	strg
Stove oil	stv
Straddle	strd
Straddle packer	SP
Straddle packer drill stem test	SP–DST
Straight	strt
Straight hole test	SHT
Straightening	stging
Strainer	stnr
Strand(ed)	Strd
Stratigraphic	strat
Strawn	Str
Streaked	stk
Streaks	stk
Striated	stri
String shot	SS
Stringer	strg
Strokes per minute	SPM
Stromatoporoid	strom
Strong	strg
Structural	struc
Structure	struc
Stuck	stk

Stuffing box	SB
Styolite	styo
Styolitic	styo
Service	svc
Service unit	svcu
Sub	sb
Sub-Clarksville	Sub Clarks
Subdivision	subd
Subsea	SS
Subsidiary	sub
Substance	sub
Substation	substa
Subsurface	SS
Sucrose	suc
Sucrosic	suc
Suction	suct
Sugary	sug
Sulphated	sulph
Sulphur (sulfur)	sul
Sulfur by bomb method	S Bomb
Sulphur water	sul wtr
Summary	sum
Summerville	Sum
Sunburst	Sb
Sunburst	Sunb
Sundance	Sund
Supai	Sup
Superintendent	supt
Superseded	supsd
Supervisor	suprv
Supplement	supp
Supply	sply
Supply (ied) (ier) (ing)	supl
Support	suppt
Surface	sfc
Surface	surf
Surface geology	SG
Surface measurement	SM
Surface pressure	SP

Surplus	surp
Survey	sur
Suspended	susp
Swab and flow	S&F
Swabbed	S/
Swabbed, swabbing	swb (d) (g)
Swabbing unit	SWU
Swage	swg
Swaged	swd
Swastika	Swas
Sweetening	swet
Switchboard	swbd
Switchgear	swgr
Sycamore	Syc
Sylvan	Syl
Symbol	sym
Symmetrical	sym
Synchronizing	syn
Synchronous	syn
Synchronous converter	syn conv
Synthetic	syn
System	sys

T

Tabular	tab
Tabulating	tab
Tagliabue	Tag
Tallahatta	Tal
Tampico	Tamp
Tank	tk
Tank battery	TB
Tank car	T/C
Tank truck	TT
Tank wagon	TW
Tankage	tkg
Tanker(s)	tkr

Tannehill	Tann
Tansill	Tan
Tarkio	Tark
Tarred and wrapped	T&W
Tar Springs sand	TS
Taste	tste
Taylor	Tay
Tearing out rotary tools	TORT
Technical	tech
Technician	tech
Tee	T
Telegraph	tel
Telegraph Creek	Tel Cr
Telephone	tel
Teletype	TWX
Television	TV
Temperature	Temp
Temperature control valve	TCV
Temperature controller	TC
Temperature differential indicator	TDI
Temperature differential recorder	TDR
Temperature indicator	TI
Temperature indicator controller	TIC
Temperature recorder	TR
Temperature recorder controller	TRC
Temperature Survey indicated top cement at	TSITC
Temporarily abandoned	TA
Temporarily shut down	TSD
Temporarily shut in	TSI
Temporary (ily)	Temp
Temporary dealer allowance	TDA
Temporary voluntary allowance	TVA
Tender	tndr
Tensile strength	TS
Tensleep	Tens
Tentaculites	Tent
Tentative	tent
Teremplealeau	Tremp
Terminal	term

Terminate (d) (ing) (ion)	termin
Tertiary	Ter
Tertiary butyl alcohol	TBA
Test (ed) (ing)	tst (d) (g)
Test to follow	TTF
Tester	tstr
Testing on pump	TOP
Tetraethyl lead	TEL
Tetramethyl lead	TML
Texana	Tex
Textularia articulate	Text. art.
Textularia dibollensis	Text. d.
Textularia hockleyensis	Text. h.
Textularia warreni	Text. w.
Texture	tex
Thaynes	Thay
Thence	th
Thermal	thrm
Thermal cracker	thrm ckr
Thermofor catalytic cracking	TCC
Thermometer	therm
Thermopolis	Ther
Thermostat	therst
Thick	thk
Thickness	thk
Thin bedded	TB
Thousand cubic feet	MCF
Thousand cubic feet per day	MCFD
Thousand British thermal units	MBTU
Thread	thd
Thread large end	TLE
Thread on both ends	TOBE
Thread small end	TSE
Thread small end, weld large end	TSE–WLE
Threaded	thd
Threaded & coupled	T & C
Threaded both ends	TBE
Threaded one end	TOE
Threaded pipe flange	TPF
Three Finger	Tfing

Three Forks	Tfks
Throttling	thrling
Through	thru
Through tubing	TT
Thurman	Thur
Tight	ti
Tight hole	TH
Tight no show	TNS
Timpas	Tim
Timpoweap	Timpo
Tires, batteries and accessories	TBA
Todilto	Tod
Tolerance	tol
Toluene	tolu
Ton (after a number)	T
Tongue and groove (joint)	T&G
Tonkawa	Tonk
Too small to measure	TSTM
Too weak to measure	TWTM
Tool	tl
Tool closed	TC
Tool open	TO
Tool pusher	TP
Tools	tl
Top and bottom	T&B
Top & bottom chokes	T&BC
Top choke	TC
Top hole flow pressure	THFP
Top of (a formation)	T/
Top of cement	TOC
Top of cement plug	TOCP
Top of fish	TOF
Top of liner	TOL
Top of liner hanger	TLH
Top of pay	T/pay
Top salt	T/S
Top of sand	T/sd
Topeka	Tpka
Topographic	topo
Topography	topo

Topping	topg
Topping and coking	T & C
Toronto	Tor
Toroweap	Toro
Total	tot
Total depth	TD
Total time lost	TTL
Totally enclosed-fan cooled	TEFC
Tough	tgh
Towanda	Tow
Township (as T2N)	T
Township	twp
Townsite	twst
Trace	tr
Trackage	trkg
Tract	tr
Transfer (ed) (ing)	trans
Transformer	trans
Translucent	transl
Transmission	trans
Transparent	transp
Transportation	transp
Travis Peak	TP
Treat (ed) (ing)	trt (d) (g)
Treater	trtr
Trenton	Tren
Trenton	Trn
Triassic	Tri
Triscresyl phosphate	TCP
Trilobite	trilo
Trinidad	Trin
Trip in hole	TIH
Trip out of hole	TOH
Triplicate	trip
Tripoli	trip
Tripolitic	trip
Tripped (ing)	trip
Truck	trk
True boiling point	TBP
True Vapor Process	TVP

True vertical depth	TVD
Tube	tb
Tubing	tbg
Tubing and rods	T&R
Tubing choke	TC
Tubing choke	tbg chk
Tubing pressure	TP
Tubing pressure	tbg press
Tubing pressure–closed	TPC
Tubing pressure–flowing	TPF
Tubing pressure shut in	TPSI
Tubinghead flange	THF
Tucker	Tuck
Tuffaceous	tfs
Tuffaceous	tuf
Tulip Creek	Tul Cr
Tungsten carbide	tung carb
Turn around	TA
Turned over to producing section	TOPS
Turned to test tank	TTTT
Turnpike	tpk
Tuscaloosa	Tus
Twin Creek	Tw Cr
Twisted off	twst off
Type	ty
Typewriter	tywr
Typical	typ

U

Ultimate	ult
Ultra high frequency	UHF
Unbranded	unbr
Unconformity	unconf
Unconsolidated	uncons
Under construction	U/C
Under digging	UD

Under gauge	UG
Under ground	UG
Under reaming	UR
Undifferentiated	undiff
Unfinished	unf
Uniform	uni
Union Valley	UV
Unit	un
Universal	univ
Universal gear lubricant	UGL
University	univ
Unsulfonated residue	UR
Upper (i.e., U/Simpson)	U/
Upper and lower	U/L
Use customer's hose	UCH
Used with	U/W
Uvigerina lirettensis	Uvig. lir.

V

Vacant	vac
Vacation	vac
Vacuum	vac
Vaginulina regina	Vag. reg
Valera	Val
Valve	vlv
Vanguard	Vang
Vapor	vap
Vapor pressure	VP
Vapor-liquid ratio	V/L
Varas	vrs
Variable	var
Variegated	vari
Various	var
Varnish makers & painters naphtha	VM&P Naphth
Varved	vrvd
Velocity	vel

Velocity survey	V/S
Ventilator	vent
Verdigris	Verd
Vermillion Cliff	Ver Cl
Versus	vs
Vertebrate	vrtb
Vertical	vert
Vertical	vrtl
Very (as very tight)	v.
Very common	v.c.
Very fine-grained	v-f-gr
Very heavily oil-cut mud	v-HOCM
Very high frequency	VHF
Very light amber cut	VLAC
Very noticeable	v.n.
Very poor sample	VPS
Very rare	v.r.
Very slight	v-sli
Very slight gas-cut mud	VSGCM
Very slight show of gas	VSSG
Very slight show of oil	VSSO
Very slightly porous	VSP
Vesicular	ves
Vicksburg	Vks
Viola	Vi
Virgelle	Virg
Viscosity	vis
Viscosity index	VI
Visible	vis
Vitreous	vit
Vitrified clay pipe	VCP
Vogtsberger	Vogts
Volt	v
Volt-ampere	va
Volt-ampere reactive	var
Volume	V
Volume	vol
Volumetric efficiency	vol. eff.
Vuggy	vug
Vugular	vug

W

Wabaunsee	Wab
Waddell	Wad
Waiting	wtg
Waiting on	WO
Waiting on acid	WOA
Waiting on allowable	WOA
Waiting on battery	WOB
Waiting on cable tools	WOCT
Waiting on cement	WOC
Waiting on completion rig	WOCR
Waiting on completion tools	WOCT
Waiting on orders	WOO
Waiting on permit	WOP
Waiting on pipe	WOP
Waiting on potential test	WOPT
Waiting on production equipment	WOPE
Waiting on pump	WOP
Waiting on pumping unit	WOPU
Waiting on rig or rotary	WOR
Waiting on rotary tools	WORT
Waiting on standard tools	WOST
Waiting on state potential	WOSP
Waiting on tank & connection	WOT&C
Waiting on test or tools	WOT
Waiting on weather	WOW
Wall (if used with pipe−	W
Wall Creek	W Cr
Wall thickness (pipe)	WT
Waltersburg sand	Wa Sd
Wapanucka	Wap
Warehouse	whse
Warsaw	War
Wasatch	Was
Wash over	WO
Wash pipe	WP
Wash water	WW

Washed	wshd
Washing	wshg
Washing in	WI
Washita	Wash
Washita-Fredericksburg	W-F
Washover string	WOS
Water, watery	wtr (y)
Water blanket	WB
Water cushion (DST)	WC
Water cushion to surface	WCTS
Water cut	WC
Water depth	WD
Water disposal well	WD
Water in hole	WIH
Water injection	WI
Water load	W/L
Water loss	WL
Water not shut-off	WNSO
Water, oil or gas	WOG
Water shut-off	WSO
Water shut-off no good	WSONG
Water shut-off ok	WSOOK
Water supply well	WSW
Water to surface	WTS
Water well	WW
Water with slight show of oil	W/SSO
Water with sulphur odor	W/sulf O
Water-cut mud	WCM
Water-cut oil	WCO
Waterflood	WF
Water-oil ratio	WOR
Watt	w
Weak	wk
Weak air blow	WAB
Weather	wthr
Weathered	wthd
Weber	Web
Week	wk
Weight	wgt
Weight	wt

Weight on bit	W.O.B.
Weld ends	WE
Weld neck	WN
Welded	wld
Welder	wldr
Welding	wld
Welding neck	WN
Wellhead	WH
Wellington	Well
Went back in hole	WBIH
Went in hole	WIH
West	W
West half	W/2
West line	WL
West Offset	W/O
Westerly	W'ly
Wet bulb	WB
Whipstock	WS
Whipstock	whip
Whipstock depth	WSD
White	wht
White Dolomite	Wh Dol
White River	WR
White sand	Wh Sd
Wholesale	whsle
Wichita	Wich.
Wichita Albany	Wich Alb
Wide flange	WF
Wilcox	Wx
Wildcat	WC
Wildcat field discovery	WFD
Willberne	Willb
Wind River	Wd R
Winfield	Winf
Wingate	Wing
Winnipeg	Winn
Winona	Win
Wire line	WL
Wire line coring	WLC
Wireline test	WLT

Wireline total depth	WLTD
With	w/
Without drill pipe	WODP
Wolfe City	WC
Wolfcamp	Wolfc
Woodbine	WB
Woodford	Wdfd
Woodford	Woodf
Woodside	Wood
Work order	WO
Worked	wkd
Working	wkg
Working interest	WI
Working pressure	WP
Workover	wko
Workover	WO
Workover rig	wkor
Wrapper	wpr
Wreford	Wref
Wrought iron	WI

X

X-ray	X-R

Y

Yard(s)	yd
Yates	Y
Yazoo	Yz
Year	yr
Yellow	yel
Yellow indicating lamp	YIL
Yield point	YP

Yoakum	Yoak
Your message of date	YMD
Your message yesterday	YMY

Z

Zenith	zen
Zilpha	Zil
Zone	Z

ABBREVIATIONS FOR LOGGING
TOOLS AND SERVICES

(The appropriate companies and associations have not yet established standard abbreviations for the logging segment of the oil and gas industry. The following lists, by individual companies, are included for convenience.)

DRESSER ATLAS

Log or Service	Abbreviation
Acoustilog	ALC
Acoustilog Caliper Gamma Ray	ALC-GR
Acoustilog Caliper Neutron	ALC-N
Acoustilog Caliper Gamma Ray-Neutron	ALC-GRN
Acoustic Cement Bond	CBL
Acoustic Cement Bond Gamma Roy	CBL-GR
Acoustic Cement Bond Neutron	CBL N
Acoustic Cement Bond G/R Neutron	CBL GRN
Acoustic Parameter—Depth	AC PAR D
Acoustic Parameter—Logging	AC PAR L
Acoustic Parameter—16 mm Scope	AC PAR 16
Acoustic Signature	AC SIGN
Atlantic Chlorinlog	A CHL
Borehole Compensated	BHC
BHC Acoustilog Caliper	BHC ALC
BHC Acoustilog Caliper Gamma Ray	BHC ALC GR
BHC Acoustilog Caliper Neutron	BHC ALC N
BHC Acoustilog Caliper G/R Neutron	BHC ALC GRN
BHC Acoustilog Caliper (Thru Casing)	BHC AL TC
BHC Acoustilog Caliper Gamma Ray (Thru Casing)	BHC AL GR TC
BHC Acoustilog Caliper G/R Neutron (Thru Casing)	BHC AL GRN TC
Caliper	CL

Casing Potential Profile	CPP
Cemotop	CTL
Channelmaster	CML
Channelmaster–Neutron	CML N
Chlorinlog	CHL
Chlorinlog–Gamma Ray	CHL GR
Compensated Densilog Caliper	C DLC
Compensated Densilog Caliper Gamma Ray	C DLC GR
Compensated Densilog Caliper Neutron	C DLC N
Compensated Densilog Caliper G/R Neutron	C DLC GRN
Compensated Densilog Caliper Minilog	C DLC M
Conductivity Derived Porosity	CDP
Corgun	CG
Densilog Caliper Gamma Ray	DLC GR
Depth Determination	DD
Directional Survey	DS
Dual Induction Focused Log	DIFL
Dual Induction Focused Log Gamma Ray	DIFL-GR
Electrolog	EL
Focused Diplog	F DIP
Formation Tester	FT
4 Arm High Resolution Diplog	R H DIP
Frac Log	FRAC L
Frac Log–Gamma Ray	FRAC-GR
Gamma Ray Cased Hole	GR CH
Gamma Ray/Dual Caliper	GR/D CALIPER
Gamma Ray–Open Hole	G/R OH
Gamma Ray Neutron Cased Hole	GRN CH
Geophone	GEO
Gamma Ray Neutron–Open Hole	GR/N OH
Induction Electrolog	IEL
Induction Electrolog Gamma Ray	IEL-GR
Induction Electrolog Neutron	IEL-N
Induction Electrolog Gamma Ray Neutron	IEL-GRN

Induction Log	IL
Induction Log–Gamma Ray	IL-GR
Induction Log Neutron	IL-N
Laterolog	LL
Laterolog-Gamma Ray	LL-GR
Laterolog-Neutron	LL-N
Laterolog-Gamma Ray-Neutron	LL-GRN
Microlaterolog-Caliper	MLLC
Minilog Caliper	ML-C
Minilog Caliper Gamma Ray	ML-C-GR
Movable Oil Plot	MOP
Nuclear Flolog	NFL
Nuclear Flolog–Gamma Ray	NFL GR
Nuclear Flolog–Neutron	NFL N
Nuclear Flolog–Gamma Ray Neutron	NFL GRN
Nuclear Cement Log	NCL
Neutron Cased Hole	N CH
Neutron Open Hole	N OH
Neutron Lifetime	NLL
Neutron Lifetime Gamma Ray	NLL GR
Neutron Lifetime Neutron	NLL N
Neutron Lifetime G/R–Neutron	NLL GRN
Neutron Lifetime CBL	NLL CBL
Neutron Lifetime–G/R–CBL	NLL GR CBL
Neutron Lifetime–CBL Neutron	NL CB N
Neutron Lifetime–CBL G/R–Neutron	NLL CBL GR N
Perforating Control	PFC
PFC Gamma Ray	PFC GR
PFC Neutron	PFC N
Photon	PL
Proximity Minilog	PROX-MLC
Sidewall Neutron	SWN
Sidewall Neutron–Gamma Ray	SWN GR
Temperature–Differential	DTL
Temperature–Gamma Ray–Neutron	T GRN
Temperature Log	TL
Temperature Log–Gamma Ray	T GR
Temperature–Neutron	T N
Total Time Integrator	TTI

Tracer Log	TL
Tracer Log—Neutron	TNL
Tracer Material	TM
Tracer Placement with Dump Bailer	TU DB
Tricore	TCS

GO INTERNATIONAL

Log or Service	Abbreviation
Caliper	CALP
Cement Bond Log	CBL
Differential Temperature	DIF-T
Gamma Ray	GR
Gamma Ray Neutron	GR-N
Neutron	N
Temperature Log	T

SCHLUMBERGER WELL SERVICES

Log or Service	Abbreviation
Amplitude Logging	A-BHC
Bore Hole Compensated	BHC
BHC Sonic Logging	BHC
BHC Sonic-Gamma Ray Logging	BHC-GR
Bridge Plug Service	BP
Borehole Televiewer	TVT
Caliper Logging	CAL
Casing Cutter Service	SCE-CC
Cement Bond Logging	CBL
Cement Bond-Gamma Ray Logging	CBL-GR
Cement Bond-Variable Density Logging	CBL-VD
Cement Dump Bailer Service	DB
Computer Processed Interpretation	MCT
Customer Instrument Service	ICS
Data Transmission	TRD
Depth Determinations	DD
Diamond Core Slicer	SS
Dipmeter-Digital	HDT-D
Directional Service	CDR

Dual Induction-Laterologging	DIL
Electric Logging	ES
Flowmeter	CFM, PFM
Formation Density Logging	FDC
Formation Density-Gamma Ray Logging	FDC-GR
Formation Testing	FT
Gamma Ray Logging	GR
Gamma Ray-Neutron Logging	GRN
Gradiomanometer	GM
High Resolution Thermometer	HRT
Induction-Gamma Ray Logging	I-GR
Induction-Electric Logging	I-ES
Junk Catcher	JB
Magnetic Taping	TPG
Microlog	ML
Neutron Logging	NL
Orienting Perforating Service	OPR
Perforating-Ceramic DPC	SCE
Perforating Depth Control	PDC
Perforating-Expendable Shaped Charge	SCE
Perforating-Hyper-Jet	SCH
Perforating-Hyper Scallop	SPH
Pressure Control	PC
Production Combination Tool Logging	PCT
Production Packer Service	PPS
Proximity-Microlog	ML
Radioactive Tracer Logging	RTP
Rwa Logging	FAL
Salt Dome Profiling	ES-ULS
Seismic Reference Service	SRS
Sidewall Coring	CST
SNP Neutron Logging	SNP
SNP Neutron-Gamma Ray Logging	SNP-GR
Synergetic Log Systems	MCT
Temperature Logging	T
Temperature-Gamma Ray Logging	T-GR
Thermal Decay Logging	TDT
Thru-Tubing Caliper	C-C
Tubing, Cutter Service	SCE-CC
Variable Density Logging	BHC-VD
Variable Density-Gamma Ray Logging	VD-GR

WELEX

Log or Service	Abbreviation
Analog Computer Service	An Cpt. Ser
Caliper Log	Cal
Compensated Acoustic Velocity Log	Com AVL
Compensated Acoustic Velocity Gamma Ray	Com AVL-G
Compensated Acoustic Velocity Neutron Log	Com AVL-N
Compensated Density Log	Com Den
Compensated Density Gamma Ray Log	Com Den-G
Computer Analyzed Logs	CAL
Contact Caliper Log	Cont
Continuous Drift Log	Con Dr.
Density Log	Den
Density Gamma Ray Log	Den-G
Depth Determination	DeDet
Digital Tape Recording Service	Dgt Tp Rec
Dip Log Digital Recording Service	Dgt Dip Rec.
Drift Log	Dr
Drill Pipe Electric Log	DPL
Electric Log	EL
Electro-Magnetic Corrosion Detector	Cor Det
Fluid Travel Log	FTrL
Formation Tester	FT
FoRxo Caliper Log	FoRxo
Frac-Finder Micro-Seismogram Log	FF-MSG
Frac-Finder Micro-Seismogram Gamma Log	FF-MSG-G
Frac-Finder Micro-Seismogram Neutron Log	FF-MSG-N
Gamma Guard Log	G-Grd
Gamma Ray Log	GR
Gamma Ray Depth Control Log	GRDC
Guard Log	Grd
High Temperature Equipment	HTEq
Induction Electric Gamma Ray Log	IEL-G
Induction Electric Neutron Log	IEL-N

Induction Electric Log	IEL
Induction Gamma Ray Log	Ind-G
Micro-Seismogram Log, Cased Hole	MSG-CBL
Micro-Seismogram Gamma Collar Log, Cased	MSG-CBL-G
Micro-Seismogram Neutron Collar Log, Cased	MSG-CBL-N
Neutron Log	NL
Neutron Depth Control Log	NDC
Precision Temperature Log	Pr Temp
Radiation Guard Log	R/A Grd
Radioactive Tracer Log	R/A Tra
Resistivity Dip Log	Dip
Side Wall Coring	SWC
Sidewall Neutron Log	SWN
Sidewall Neutron Gamma Ray Log	SWN-G
Simultaneous Gamma Ray Neutron	GRN
Special Instrument Service	Sp Inst Ser
True Vertical Depth	TVD

AAODC	See IADC
AAPG	American Association of Petroleum Geologists
AAPL	American Association of Petroleum Landmen
ACS	American Chemical Society
ADDC	Association of Desk and Derrick Clubs of North America
AEC	Atomic Energy Commission
AECRB	Alberta Energy Conservation Resources Board
AGA	American Gas Association
AGI	American Geological Institute
AGTL	Alberta Gas Trunkline Co., Ltd.
AGU	American Geophysical Union
AIChE	American Institute of Chemical Engineers
AIME	American Institute of Mining, Metallurgical and Petroleum Engineers
AISI	American Iron and Steel Institute
ANSI	American National Standards Institute
AOCS	American Oil Chemists Society
AOPL	Association of Oil Pipe Lines
AOSC	Association of Oilwell Servicing Contractors
API	American Petroleum Institute
APRA	American Petroleum Refiners Association
APW	Association of Petroleum Writers
ARCO	Atlantic Richfield Co.
ARKLA	Arkansas Louisiana Gas Co.
ASCE	American Society of Civil Engineers
ASHRAE	American Society of Heating, Refrigerating, and Air-Conditioning Engineers, Inc.

ASLE	American Society of Lubricating Engineers
ASME	American Society of Mechanical Engineers
ASPG	American Society of Professional Geologists
ASSE	American Society of Safety Engineers
ASTM	American Society for Testing Materials
AWS	American Welding Society
BLM	Bureau of Land Management
BLS	Bureau of Labor Statistics
BP	British Petroleum
BuMines	Bureau of Mines, U.S. Department of the Interior
CAGC	A combine: Continental Oil Co., Atlantic Richfield Co., Getty Oil Co., and Cities Service Oil Co.
CAODS	Canadian Association of Oilwell Drilling Contractors
CCCOP	Conservation Committee of California Oil Producers
CDS	Canadian Development Corp.
CFR	Coordinating Fuel Research Committee
GA	Canadian Gas Association
CGA	Clean Gulf Associates
CGTC	Columbia Gas Transmission Corp.
CONOCO	Continental Oil Co.
CORCO	Commonwealth Oil Refining Co., Inc.
CORS	Canadian Operational Research Society
CPA	Canadian Petroleum Association
CRC	Coordinating Research Council, Inc.
DOT	Department of Transportation
EMR	Department of Energy, Mines, and Resources (Canada)
EPA	Environmental Protection Agency

ERCB	Energy Resource Conservation Board (Alberta, Canada)
FAA	Federal Aviation Agency
FCC	Federal Communications Commission
FPC	Federal Power Commission
FTC	Federal Trade Commission
GAMA	Gas Appliance Manufacturers Association
GNEC	General Nuclear Engineering Co.
IADC	International Association of Drilling Contractors (formerly AAODC)
IAE	Institute of Automotive Engineers
ICC	Interstate Commerce Commission
IEEE	Institute of Electrical and Electronics Engineers
IGT	Institute of Gas Technology
INGAA	Independent Natural Gas Association of America
IOCA	Independent Oil Compounders Association
IOCC	Interstate Oil Compact Commission
IOSA	International Oil Scouts Association
IP	Institute of Petroleum
IPAA	Independent Petroleum Association of America
IPAC	Independent Petroleum Association of Canada
IPE	International Petroleum Exposition
IPP/L	Interprovincial Pipe Line Co.
IRAA	Independent Refiners Association of America
ISA	Instrument Society of America
KERMAC	Kerr-McGee Corp.
KIOGA	Kansas Independent Oil and Gas Association
LL&E	Louisiana Land & Exploration Co.
MIOP	Mandatory Oil Import Program
NACE	National Association of Corrosion Engineers

NACOPS	National Advisory Committee on Petroleum Statistics (Canada)
NAS	National Academy of Science
NASA	National Aeronautical and Space Administration
NEB	National Energy Board (Canada)
NEMA	National Electrical Manufacturers Association
NEPA	National Environmental Policy Act of 1969
NGPA	Natural Gas Processors Association
NGPSA	Natural Gas Processors Suppliers Association
NLGI	National Lubricating Grease Institute
NLPGA	National Liquefied Petroleum Gas Association
NLRB	National Labor Relations Board
NOFI	National Oil Fuel Institute
NOIA	National Ocean Industries Association
NOJC	National Oil Jobbers Council
NOMADS	National Oil-Equipment Manufacturers and Delegates Society
NPC	National Petroleum Council
NPRA	National Petroleum Refiners Association
NSF	National Science Foundation
OCR	Office of Coal Research
OEP	Office of Emergency Preparedness
OIA	Oil Import Administration
OIAB	Oil Import Appeals Board
OIC	Oil Information Committee
OIPA	Oklahoma Independent Petroleum Association
OOC	Offshore Operators Committee
OPC	Oil Policy Committee
OXY	Occidental Petroleum Corp.
PAD	Petroleum Administration for Defense
PESA	Petroleum Equipment Suppliers Association

PETCO	Petroleum Corporation of Texas
PGCOA	Pennsylvania Grade Crude Oil Association
PIEA	Petroleum Industry Electrical Association
Plato	Pennzoil Louisiana and Texas Offshore
PLCA	Pipe Line Contractors Association
POGO	Pennzoil Offshore Gas Operators
PPI	Plastic Pipe Institute
PPROA	Panhandle Producers and Royalty Owners Association
RMOGA	Rocky Mountain Oil and Gas Association
R-PAT	Regional Petroleum Associations of Texas
SACROC	Scurry Area Canyon Reef Operators Committee
SAE	Society of Automotive Engineers
SEG	Society of Exploration Geophysicists
SEPM	Society of Economic Paleontologists and Mineralogists
SGA	Southern Gas Association
SLAM	A combine: Signal Oil and Gas Co., Louisiana Land & Exploration Co., Amerada Hess Corp., and Marathon Oil Co.
SOCAL	Standard Oil Company of California
SOHIO	Standard Oil Co. of Ohio
SPE	Society of Petroleum Engineers of AIME
SPEE	Society of Petroleum Evaluation Engineers
SPWLA	Society of Professional Well Log Analysts
STATCAN	Statistics Canada ex Dominion Bureau of Statistics (DBS)
TCP	Trans-Canada Pipe Lines Ltd.
TETCO	Texas Eastern Transmission Corp.
TGT	Tennessee Gas Transmission Co.

THUMS	A combine: Texaco, Inc., Humble Oil & Refining Co., Union Oil Co. of California, Mobil Oil Corp., and Shell Oil Co.
TIPRO	Texas Independent Producers and Royalty Owners Association
TRANSCO	Transcontinental Gas Pipe Line Corp.
USGS	United States Geological Survey
WeCTOGA	West Central Texas Oil and Gas Association
WPC	World Petroleum Congress

COMPANIES AND ASSOCIATIONS
Outside the U.S. and Canada

AAOC	American Asiatic Oil Corp. (Philippines)
ABCD	Asfalti Bitumi Cementi Derivati, S.A. (Italy)
ACNA	Aziende Colori Nazionali Affini (Italy)
A.C.P.H.A.	Association Cooperative pour la Recherche et l'Exploration des Hydrocarbures en Algerie (Algeria)
ADCO-HH	African Drilling Co.–H. Hamouda (Libya)
AGIP S.p.A.	Azienda Generale Italiana Petroli S.p.A. (Italy)
A.H.I. BAU	Allegemeine Hoch-und Ingenieurbau AG (Germany)
AIOC	American International Oil Co. (U.S.A.)
AITASA	Aguas Industriales de Tarragona, S.A. (Spain)
AK CHEMI	GmbH & Co. KG–subsidiary of Associated Octel, Ltd., London, Eng. (Germany)
AKU	Algemene Kunstzijde Unie, N.V. (Netherlands)

ALFOR	Societe Algerienne de Forage (Algeria)
ALGECO	Alliance & Gestion Commerciale (France)
A.L.O.R.	Australian Lubricating Oil Refinery Ltd. (Australia)
AMATEX	Amsterdamsch Tankopslagbedrijf N.V. (Netherlands)
AMI	Ausonia Mineraria (Italy)
AMIF	Ausonia Miniere Francaise (France)
AMINOIL	American Independent Oil Co. (U.S.A.)
AMOSEAS	American Overseas Petroleum Co., Ltd. (Libya)
AMOSPAIN	American Overseas Petroleum Ltd. (Spain)
ANCAP	Administracion Nacional de Combustibles, Alcohol y Portland (Uraguay)
ANIC	Azienda Nazionale Idrogenazione Combustibili S.p.A. (Italy)
AOC	Aramco Overseas Co. (Switzerland, Netherlands, Japan)
APC	Azote et Produits Chimiques (France)
APEX	American Petrofina Exploration Co. (Spain)
API	Anonima Petroli Italiana (Italy)
AQUITAINE	Societe Nationale des Petroles D'Aquitaine (France)
ARGAS	Arabian Geophysical & Surveying Co. (Saudi Arabia)
ARAMCO	Arabian American Oil Co.
ASCOP	Association Cooperative pour la Recherche et l'Exploration des Hydrocarbures en Algerie (Algeria)
ASED	Amoniaque Synthetique et Derives (Belgium)
ASESA	Alfaltos Espanoles S.A. (Spain)

ATAS	Anadolu Tastiyehanesi A.S. (Turkey)
AUXERAP	Societe Auxiliare de la Regie Autonome des Petroles (France)
AZOLACQ	Societe Chimique d'Engrais et de Produits de Synthese (France)
BAPCO	Bahrain Petroleum Co. Ltd. (Bahrain)
BASF	Badische Anilin & Soda-Frabrik AB (Germany)
BASUCOL	Barranquilla Supply & Co. (Colombia)
B.I.P.M.	Bataafse Internationale Petroleum Mij. N.V. (Netherlands)
BOGOC	Bolivian Gulf Oil Co. (Bolivia)
BORCO	Bahamas Oil Refining Co. (Bahamas)
BP	British Petroleum Co., Ltd. (England)
BRGG	Bureau de Recherches Geologique et Geophysique (France)
BRGM	Bureau de Recherches Geologiques et Minieres (France)
BRIGITTA	Gewerkschaft Brigitta (Germany)
BRP	Bureau du Recherche de Petrole (France)
BRPM	Bureau de Recherches et de Participations Mineres (Morocco)
CALSPAIN	California Oil Co. of Spain (Spain)
CALTEX	Various affiliates of Texaco Inc. and Std. of Calif.
CALVO SOTELO	Empresa Nacional Calvo Sotelo (Spain)
CAMEL	Campagnie Algerienne du Methane Liquide (France, Algeria)
CAMPSA	Compania Arrendataria del Monopolio de Petroleos, S.A. (Spain)

CAPAG	Enterprise Moderne de Canalisations Petrolieres, Aquiferes et Gazieres (France)
CARBESA	Carbon Black Espanola, S.A. (Conoco affiliate)
CAREP	Compagnie Algerienne de Recherche et d'Exploitation Petrolieres (Algiers)
CCC	Compania Carbonos Coloidais (Brazil)
C.E.C.A.	Carbonisation et Charbons Actifs S.A. (France)
CEICO	Central Espanol Ingenieria y Control S.A. (Spain)
CEL	Central European Pipeline (Germany)
CEOA	Centre Europe de'Exploitation de l'OTAN (France)
CEP	Compagnie D'Exploration Petroliere (France)
CEPSA	Compania Espanola de Petroleos, S.A. (Spain)
CETRA	Compagnie Europeanne de Canalisations et de Travaux (France)
CFEM	Compagnie Francaise d'Enterprises Metalliques (France)
CFM	Compagnie Francaise du Methane (France)
CFMK	Compagnie Ferguson Morrison-Knudsen (France)
CFP	Compagnie Francaise du Petroles (France)
CFPA	Compagnie Francaise des Petroles (Algeria) (France)
CFPS	Compagnie Francaise de Prospection Sismique (France)
CFR	Compagnie Francaise de Raffinage (France)
CGG	Compagnie Generale de Geophysique (France, Australia, Singapore)

CIAGO	N.V. Chemische Industrie aku-Goodrich (Netherlands)
CIEPSA	Compania de Investigacion y Explotaciones Petroliferas, S.A. (Spain)
CIM	Compagnie Industrielle Maritime (France)
CIMI	Compania Italiana Montaggi Industriali S.p.a. (Italy)
CINSA	Compania Insular del Nitrogena, S.A. (Spain)
CIPAO	Compagnie Industrielle des Petroles de l'A.O. (France)
CIPSA	Compania Iberica de Prospecciones, S.A. (Spain)
CIRES	Compania Industrial de Resinas Sinteticas (Portugal)
CLASA	Carburanti Lubrificanti Affini S.p.A. (Italy)
CMF	Construzioni Metalliche Finsider S.p.A. (Italy)
COCHIME	Compagnie Chimique de la Meterranee (France)
CODI	Colombianos Distribuidores de Combustibles S.A. (Colombia)
COFIREP	Compagnie Financiere de Recherches Petrolieres (France)
COFOR	Compagnie Generale de Forages (France)
COLCITO	Colombia-Cities Service Petroleum Corp. (Colombia)
COLPET	Colombian Petroleum Co. (Colombia)
COMEX	Compagnie Maritime d'Expertises (France)
CONSPAIN	Continental Oil Co. of Spain (Spain) Conoco Espanola S.A. (Spain)
COPAREX	Compagnie de Participations, de Recherches et D'Exploitations Petrolieres (France)

COPE	Compagnie Orientale des Petroles d'Egypte (Egypt)
COPEBRAS	Compania Petroquimica Brasileira (Brazil)
COPEFA	Compagnie des Petroles France-Afrique (France)
COPETAO	Compagnie des Petroles Total (Afrique Quest) (France)
COPETMA	Compagnie les Petroles Total (Madagascar)
COPISA	Compania Petrolifera Iberica, Sociedad Anonima (Spain)
COPOSEP	Compagnie des Petroles du Sud est Parisien (France)
C.O.R.I.	Compania Richerche Idrocarburi S.p.A. (Italy)
COS	Coordinated Oil Services (France)
CPA	Compagnie des Petroles d'Algerie (Algeria)
CPC	Chinese Petroleum Corporation, Taiwan, China
CPTL	Compagnie des Petroles Total (Libye) (France)
CRAN	Compagnie de Raffinage en Afrique du Nord (Algeria)
CREPS	Compagnie de Recherches et d'Exploitation de Petrole au Sahara (Algeria)
CRR	Compagnie Rhenane de Raffinage (France)
CSRPG	Chambre Syndicale de la Recherche et de la Production du Petrole et du Gaz Naturel (France)
CTIP	Compania Tecnica Industrie Petroli S.p.a. (Italy)
CVP	Corporacion Venezolano del Petroleo (Venezuela)
DCEA	Direction Centrale des Essences des Armees (France)
DEA	Deutsche Erdol-Aktiengesellschaft (Germany)

DEMINEX	Deutsch Erdolversorgungsgesellschaft mbH (Germany) (Trinidad)
DIAMEX	Diamond Chemicals de Mexico, S.A. de C.V. (Mexico)
DICA	Direction des Carburants (France)
DICA	Distilleria Italiana Carburanti Affini (Italy)
DITTA	Macchia Averardo (Italy)
DUPETCO	Dubai Petroleum Company (Trucial States)
E.A.O.R.	East African Oil Refineries, Ltd. (Kenya)
ECF	Essences et Carbutants de France (France)
ECOPETROL	Empresa Colombiana de Petroleos (Colombia)
EGTA	Enterprises et Grands Travaux de l'Atlantique (France)
ELF-ERAP	Enterprise de Recherches et d'Activites Petrolieres (France)
ELF-U.I.P.	Elf Union Industrielle des Petroles (France)
ELF-SPAFE	Elf des Petroles D'Afrique Equatoriale (France)
ELGI	M. All. Eotvos Lorand Geofizikai Intezet (Hungary)
ENAP	Empresa Nacional del Petroleo (Chile)
ENCAL	Engenheiros Consultores Associados S.A. (Brazil)
ENCASO	Empresa Nacional Calvo Sotelo de Combustibles Liquidos y Lubricantes, S.A. (Spain)
ENGEBRAS	Engenharia Especializada Brasileira, S.A. (Brazil) (Venezuela)
ENI	Ente Nazionale Idrocarburi (Italy)
ENPASA	Empresa Nacional de Petroleos de Aragon, S.A. (Spain)
ENPENSA	Empresa Nacional de Petroleos de Navarra, S.A. (Spain)

ERAP	Enterprise de Recherches et d'Activites Petrolieres (France)
ESSAF	Esso Standard Societe Anonyme Francaise (France)
ESSOPETROL	Esso Petroleos Espanoles, S.A. (Spain)
ESSO REP	Societe Esso de Recherches et Exploitation Petrolieres (France)
E.T.P.M.	Societe Entrepose G.T.M. pour les Travaux Petroliers Maritimes (France)
EURAFREP	Societe de Recherches et D'Exploitation de Petrole (France)
FERTIBERIA	Fertilizantes de Iberia, S.A. (Spain)
FFC	Federation Francaise des Carburants (France)
FINAREP	Societe Financiere des Petroles (France)
FOREX	Societe Forex Forages et Exploitation Petrolieres (United Kingdom)
FRANCAREP	Compagnie Franco-Africaine de Recherches Petrolieres (France)
FRAP	Societe de Construction de Feeders, Raffineries, Adductions d'Eau et Pipe-Lines (France)
FRISIA	Erdolwerke Frisia A.G. (Germany)
GARRONE	Garrone (Dott. Edoardo) Raffineria Petroli S.a.S. (Italy)
GBAG	Gelsenberg Benzin (Germany)
GESCO	General Engineering Services (Colombia)
GHAIP	Ghanian Italian Petroleum Co., Ltd. (Ghana)
GPC	The General Petroleum Co. (Egypt)
G.T.M.	Les Grands Travaux de Marseille (France)
HELIECUADOR	Helicopteros Nacionales S.A. (Ecuador)
HDC	Hoechst Dyes & Chemicals Ltd. (India)

HIDECA	Hidrocarburos y Derivados C.A. (Brazil, Uraguay & Venezuela)
HIP	Hemijska Industrija Pancevo (Yugoslavia)
H.I.S.A.	Herramientas Interamericanas, S.A. de C.V. (Mexico)
HISPANOIL	Hispanica de Petroleos, S.A. (Spain)
HOC	Hindustan Organic Chemicals Ltd. (India)
HYLSA	Hojalata Y Lamina, S.A. (Mexico)
IAP	Institut Algerien du Petrole (Algeria)
ICI	Imperial Chemical Industries Ltd. (England)
ICIANZ	Imperial Chemical Industries of Australia & New Zealand Ltd. (Australia, New Zealand)
ICIP	Industrie Chimiche Italiane del Petrolio (Italy)
IEOC	International Egyptian Oil Co., Inc. (Egypt)
IFCE	Institut Francais des Combustibles et de l'Energie (France)
IFP	Institute Francaise du Petrole (France)
IGSA	Investigaciones Geologicas, S.A. (Spain)
IIAPCO	Independent Indonesian American Petroleum Co. (Indonesia)
I.L.S.E.A.	Industria Leganti Stradali et Affini (United Kingdom)
I.M.E.	Industrias Matarazzo de Energia (Brazil)
IMEG	International Management & Engineering Group of Britain Ltd. (United Kingdom)
IMEG	Iranian Management & Engineering Group Ltd. (Iran)

IMINOCO	Iranian Marine International Oil Co. (Iran)
IMS	Industria Metalurgica de Salvador, S/Z (Brazil)
I.N.C.I.S.A.	Impresa Nazionale Condotte Industriali Strade Affini (United Kingdom)
INDEIN	Ingenieria Y Desarrolio Industrial S.A. (Spain)
INI	Instituto Nacional de Industria (Spain)
INOC	Iraq National Oil Co. (Iraq)
INTERCOL	International Petroleum (Colombia) Ltd. (Colombia)
IODRIC	International Oceanic Development Research Information Center (Japan)
IOE & PC	Iranian Oil Exploration & Producing Co. (Iran)
IORC	Iranian Oil Refining Co. N.V. (United Kingdom)
IPAC	Iran Pan American Oil Co. (Iran)
I.P.L.O.M.	Industria Piemontese Lavorazione Oil Minerali (United Kingdom)
IPLAS	Industrija Plastike (Yugoslavia)
IPRAS	Instanbul Petrol Rafinerisi A.S. (Turkey)
IRANOP	Iranian Oil Participants Limited (England)
IROM	Industria Raffinazione Oil Minerali (Italy)
IROPCO	Iranian Offshore Petroleum Company (United Kingdom)
IROS	Iranian Oil Services, Ltd. (England)
IVP	Instituto Venezolano de Petroquimica (Venezuela)
JAPEX	Japan Petroleum Trading Co. Ltd. (Japan)
KIZ	Kemijska Industrijska Zajednica (Yugoslavia)

KNPC	Kuwait National Petroleum Co. (Arabia)
KSEPL	Kon./Shell Exploration and Production Laboratory (Netherlands)
KSPC	Kuwait Spanish Petroleum Co. (Kuwait)
KUOCO	Kuwait Oil Co., Ltd. (England)
LAPCO	Lavan Petroleum Co. (Iran)
LEMIGAS	Lembaga Minjak Dan Gas Bumi (Indonesia)
LINOCO	Libyan National Oil Corp. (Libya)
L.M.B.H.	Lemgaga Kebajoran & Gas Bumi (Libya)
MABANAFT	Marquard & Bahls B.m.b.H. (Germany)
MATEP	Materials Tecnicos de Petroleo S.A. (Brazil)
MAWAG	Mineraloel Aktien Gesellschaft ag (Germany)
MEDRECO	Mediterranean Refining Co. (Lebanon)
MEKOG	N.B. Maatschappij Tot Exploitatie van Kooksovengassen (Netherlands) (Netherlands)
MENEG	Mene Grande Oil Co. (Venezuela)
METG	Mittelrheinische Ergastransport GmbH (Germany)
M.I.T.I.	Ministry of International Trade and Industry (Japan)
MODEC	Mitsui Ocean Development & Engineering Co. Ltd. (Japan)
MPL	Murco Petroleum Limited (England)
NAKI	Nagynyomasu Kiserleti Intezet (Hungary)
NAM	N.V. Nederlandse Aardolie Mij. (Netherlands)
NAPM	N.V. Nederlands Amerikaanse Pijpleiding Maatschappij (Netherlands)

NCM	Nederlandse Constructiebedrijven en Machinefrabriken N.V. (Netherlands)
NDSM	Nederlandsche Dok en Scheepsbouw Maatschappij (Netherlands)
NED.	North Sea Diving Services, N.V. (Netherlands)
NEPTUNE	Soc. de Forages en Mer Neptune (France)
NETG	Nordrheinische Erdgastransport Gesselschaft mbH (Germany)
NEVIKI	Nehezvegyipari Kutato Intezet (Hungary)
NIOC	National Iranian Oil Co. (Iran)
NORDIVE	North Sea Diving Services Ltd. (United Kingdom)
NOSODECO	North Sumatra Oil Development Cooperation Co. Ltd. (Indonesia)
NPC	Nederlandse Pijpleiding Constructie Combinatie (Netherlands)
NPCI	National Petroleum Co. of Iran (Iran)
N.V.A.I.G.B.	N.V. Algemene Internationale Gasleidingen Bouw (Netherlands)
N.V.G.	Nordsee Versorgungsschiffahrt GmbH (Germany)
NWO	Nord-West Oelleitung GmbH (Germany)
OCCR	Office Central de Chauffe Rationnelle (France)
OEA	Operaciones Especiales Argentinas (Argentina)
OKI	Organsko Kenijska Industrija (Yugoslavia)
OMNIREX	Omnium de Recherches et Exploitations Petrolieres (France)
OMV	Oesterreichische Mineraloelverwaltung A.G. (Austria)

OPEC	Organization of Petroleum Exporting Countries
OTP	Omnium Techniques des Transports par Pipelines (France)
PCRB	Compagnie des Produits Chimiques et Raffineries de Berre (France)
PEMEX	Petroleos Mexicanos (Mexico)
PERMAGO	Perforaciones Marinas del Golfo S.A. (Mexico)
PETRANGOL	Companhia de Petroleos de Angola (Angola)
PETRESA	Petroquimica Espanola, S.A. (Spain)
PETROBRAS	Petroleo Brasileiro S.A. (Brazil)
PETROLIBER	Compania Iberica Refinadora de Petroleos, S.A. (Spain)
PETROMIN	General Petroleum and Mineral Organization (Saudi Arabia)
PETRONOR	Refineria de Petroleos del Norte, S. A. (Spain)
PETROPAR	Societe de Participations Petrolieres (France)
PETROREP	Societe Petroliere de Recherches Dans La Region Parisienne (France)
POLICOLSA	Poliolefinas Colombianas S.A. (Colombia)
PREPA	Societe de Prospection et Exploitations Petrolieres en Alsace (France)
PRODESA	Productos de Estireno S.A. de C.V. (Mexico)
PROTEXA	Construcciones Protexa, S.A. de C.V. (Mexico)
PYDESA	Petroleos y Derivados, S.A. (Spain)
QUIMAR	Quimica del Mar, S.A. (Mexico)
RAP	Regie Autonome des Petroles (France)

RASIOM	Raffinerie Siciliane Olii Minerali (Esso Standard Italiana S.p.A.) (Italy)
RDM	De Rotterdamsche Droogdok Mij. N.V. (Netherlands)
RDO	Rhein-Donau-Oelleitung GmbH (Germany)
REDCO	Rehabilitation, Engineering and Development Co. (Indonesia)
REPESA	Refineria de Petroleos de Escombreras, S.A. (Spain)
REPGA	Recherche et Exploitation de Petrole et de Gaz (France)
RIOGULF	Rio Gulf de Petroleos, S.A. (Spain)
SACOR	Sociedade Anonima Concessionaria da Refinacao de Petroleos em Portugal (Portugal)
SAEL	Sociedad Anonima Espanola de Lubricantes (Spain)
S.A.F.C.O.	Saudi Arabian Refinery Co. (Saudi Arabia)
SAFREP	Societe Anonyme Francaise de Recherches et D'Exploitation de Petrole (France)
SAIC	Sociedad Anonima Industrial y Commercial (Argentina)
SAM	Societe d'Approvis de Material Patrolier (France)
SAP	Societe Africaine des Petroles (France)
SAPPRO	Societe Anonyme de Pipeline a Produits Petroliers sur Territoire Genevois (Switzerland)
SAR	Societe Africaine de Raffinage (Dakar)
SARAS	S.p.a. Raffinerie Sarde (Italy)
SARL	Chimie Development International (Germany)
SAROC	Saudi Arabia Refinery Co. (Saudi Arabia)

SAROM	Societa Azionaria Raffinazione Olii Minerali (Italy)
SARPOM	Societa per Azioni Raffineria Padana Olii Minerali (Italy)
SASOL	South African Coal, Oil and Gas Corp. Ltd. (South Africa)
S.A.V.A.	Societa Alluminio Veneto per Azioni (Italy)
SCC	Societe Chimiques des Charbonnages (France)
SCI	Societe Chimie Industrielle (France)
SCP	Societe Cherifienne des Petroles (Morocco)
SECA	Societe Europreeme des Carburants (Belgium)
SEHR	Societe d'Exploitation des Hydrocarbures d'Hassi R'Mel (France)
SEPE	Sociedad de Exploracion de Petroleos Espanoles, S. A. (Spain)
SER	Societe Equatoriale de Raffinage (Gabon)
SERCOP	Societe Egyptienne pour le Raffinage et le Commerce du Petrole (Egypt)
SEREPT	Societe de Recherches et D'Exploitation des Petroles en Tunisia (Tunisia)
SER VIPETROL	Transportes Y Servicios Petroleros (Ecuador)
SETRAPEM	Societe Equatoriale de Travaux Petroliers Maritimes (France, Germany)
SFPLJ	Societe Francaise de Pipe Line du Jura (France)
SHELLREX	Societe Shell de Recherches et D'Exploitations (France)
S.I.B.P.	Societe Industrielle Belge des Petroles (Belgium)

SIF	Societe Tunisienne de Sondages, Injections, Forages (Tunisia)
SINCAT	Societa Industriale Cantese S.p.a. (Italy)
SIPSA	Sociedad Investigadora Petrolifera S.A. (Spain)
SIR	Societa Italiana Resine (Italy)
SIREP	Societe Independante de Recherches et d'Exploitation du Petrole (France)
SIRIP	Societe Irano-Italienne des Petroles (Iran)
SITEP	Societe Italo-Tunisienne d'Exploitation (Italy, Tunisia)
SMF	Societe de Fabrication de Material de Forage (France)
SMP	Svenska Murco Petroleum Aktiebolag (Sweden)
SMR	Societe Malagache de Raffinage (Malagasy)
SNGSO	Societe Nationale des Gas de Sud-Ouest (France)
SN MAREP	Societe Nationale de Material pour la Recherche et l'Exploitation du Petrole (France)
SNPA	Societe Nationale des Petroles d'Aquitaine (France)
SN REPAL	Societe Nationale de Recherches et d'Exploitation des Petroles en Algerie (France)
SOCABU	Societe du Caoutchouc Butyl (France)
SOCEA	Societe Eau et Assainissement (France
SOCIR	Societe Congo-Italienne de Raffinage (Congo Republic)
SOFEI	Societe Francaise d'Enterprises Industrielles (France)
SOGARES	Societe Gabonaise de Realisation de Structures (France)

SOMALGAZ	Societe Mixte Algerienne de Gaz (Algeria)
SOMASER	Societe Maritime de Service (France)
SONAP	Sociedade Nacional de Petroleos S.A.R.I. (Portugal)
SONAREP	Sociedade Nacional de Refinacao de Petroleos S.A.R.L. (Mozambique)
SONATRACH	Societe Nationale de Transport et de Commercialisation des Hydrocarbures (Algeria, France)
SONPETROL	Sondeos Petroliferos S.A. (Spain)
SOPEFAL	Societe Petroliere Francaise en Algeria (Algeria)
SOPEG	Societe Petroliere de Gerance (France)
SOREX	Societe de Recherches et d'Exploitations Petrolieres (France)
SOTEI	Societe Tunisienne de Enterprises Industrielles (Tunisia)
SOTHRA	Societe de Transport du Gaz Naturel D'Hassi-er-r'mel a Arzew (Algeria)
SPAFE	Societe des Petroles d'Afrique Equatoriale (France)
SPANGOC	Spanish Gulf Oil Co. (Spain)
SPEICHIM	Societe Pour l'Equipment des Industries, Chimiques (France)
SPG	Societe des Petroles de la Garonne (France)
S.P.I.	Societa Petrolifera Italiana (Italy)
SPIC	Southern Petrochemical Industries Corporation Ltd.
SPLSE	Societe du Pipe-Line Sud Europeen (France)
SPM	Societe des Petroles de Madagascar (France)
SPV	Societe des Petroles de Valence (France)

SSRP	Societe Saharienne de Recherches Petrolieres (France)
STEG	Societe Tunisienne d'Electricite et de Gaz (Tunisia)
STIR	Societe Tuniso-Italienne de Raffinage (Tunisia)
TAL	Deutsche Transalpine Oelleitung GmbH (Germany)
TAMSA	Tubos de Acero de Mexico, S.A. (Mexico)
TATSA	Tanques de Acero Trinity, S.A. (Mexico)
TECHINT	Compania Technica Internacional (Brazil)
TECHNIP	Compagnie Francaise d'Etudes et de Construction Technip (France)
TEXSPAIN	Texaco (Spain) Inc. (Spain)
TORC	Thai Oil Refinery Co. (Thailand)
T.P.A.O.	Turkiye Petrolleri A.O. (United Kingdom)
TRAPIL	Societe des Transports Petroliers Par Pipeline (France)
TRAPSA	Compagnie des Transports par Pipe-Line au Sahara (Algeria)
UCSIP	Union des Chambres Syndicales de l'Industrie du Petrole (France)
UGP	Union Generale des Petroles (France)
UIE	Union Industrielle et d'Enterprise (France)
UNIAO	Refinaria e Exploracao de Petroleo "UNIAO" S.A. (Brazil)
URAG	Unterweser Reederei GmbH (Germany)
URG	Societe pour l'Utilisation Rationnelle des Gaz (France)
WEPCO	Western Desert Operating Petroleum Company (Egypt)
YPF	Yacimientos Petroliferos Fiscales (Argentina)
YPFB	Yacimientos Petroliferos Fiscales Bolivianos (Bolivia)

API STANDARD OIL-MAPPING SYMBOLS

Location . . . . ○

Abandoned location . . . . erase symbol

Dry hole . . . . ⬡

Oil well . . . . ●

Abandoned oil well . . . . ●

Gas well . . . . ☼

Abandoned gas well . . . . ☼

Distillate well . . . . ◐

Abandoned distillate well . . . . ◐

Dual completion—oil . . . . ◉

Dual completion—gas . . . . ⊕

Drilled water-input well . . . . ⌀ W

Converted water-input well . . . . ● W

Drilled gas-input well . . . . ⌀ G

Converted gas-input well . . . . ● G

Bottom-hole location . . . . ○--–·x
   (x indicates bottom of hole. Changes in well
       status should be indicated as in symbols
       above.)

Salt-water disposal well . . . . ⊕ SWD

Courtesy American Petroleum Institute, Division of Production.

MATHEMATICAL SYMBOLS AND SIGNS

+	plus	$\therefore$	therefore
−	minus	$\because$	because
±	plus or minus	:	is to; divided by
×	multiplied by	::	as; equals
·	multiplied by	$\vdots\vdots$	geometrical proportion
÷	divided by	∝	varies as
/	divided by	$\doteq$	approaches a limit
=	equal to	∞	infinity
≠	not equal to	∫	integral
≈	nearly equal to	d	differential
≅	congruent to	∂	partial differential
≡	identical with	Σ	summation of
≢	not identical with	!	factorial product
⟷	equivalent to	π	pi (3.1416)
>	greater than	e	epsilon (2.7183)
≯	not greater than	°	degree
<	less than	′	minute; prime
≮	not less than	″	second
≧	greater than or equal to	∠	angle
≦	less than or equal to	∟	right angle
∼	difference between	⊥	perpendicular
≏	difference between	○	circle
-:	difference between	⌒	arc
√	square root	△	triangle
∛	cube root	□	square
ⁿ√	n th root	▭	rectangle

226

GREEK ALPHABET

A	α	Alpha	N	ν	Nu
B	β	Beta	Ξ	ξ	Xi
Γ	γ	Gamma	O	o	Omicron
Δ	δ	Delta	Π	π	Pi
E	ϵ	Epsilon	P	ρ	Rho
Z	ζ	Zeta	Σ	σ	Sigma
H	η	Eta	T	τ	Tau
Θ	θ	Theta	Υ	υ	Upsilon
I	ι	Iota	Φ	ϕ	Phi
K	κ	Kappa	X	χ	Chi
Λ	λ	Lambda	Ψ	ψ	Psi
M	μ	Mu	Ω	ω	Omega

METRIC-ENGLISH SYSTEMS
CONVERSION FACTORS

Basic Dimensions

Metric System

Length— meter (m)
kilometer (km)
centimeter (cm)
millimeter (mm)

Area— square meters (m²)
square centimeters (cm²)

Volume—cubic meters (m³)
cubic centimeters (cm³)
liters (l)
milliliters (ml)

Mass— Kilograms (kg)
grams (g)
gram-moles (gm-moles)
kilogram-moles (kg-moles)

Density—kg/m³, g/cm³

English System

Length— inch (in.)
foot (ft)
yard (yd)
mile (mile)

Area— square inches (in.²)
square feet (ft²)
square miles (miles²)

Volume—cubic inches (in.³)
cubic feet (ft³)
barrels (bbl)
U.S. gallons (gal)
Imperial gallons (Imp. gal)

Mass— pounds (lb)
pound-moles (lb-moles)

Density—pounds per gallon (lb/gal) lb/ft³

System Equivalents

Metric System

Length— 1 m = 100 cm = 1,000 mm = 0.001 km

Area— 1 m^2 = 10,000 cm^2

Volume—1 m^3 = 1,000,000 cm^3 = 1,000,000 ml = 1,000 l

Mass— 1 kg = 1,000 g
1 kg-mole = 1,000 gm-moles

Density—1 kg/m^3 = 0.001 g/cm^3

English System

Length— 1 ft = 12. in. = 0.333 yd = 0.000189 miles
1 mile = 5,280 ft = 1,750 yd

Area— 1 ft^2 = 144 in.^{2}
1 mile^2 = 27,878,400 ft^2

Volume—1 ft^3 = 1,728 in.^{3} = 0.178 bbl = 7.48 U.S. gal
= 6.23 Imp. gal
1 bbl = 5.61 ft^3 = 42 U.S. gal = 34.97 Imp. gal

Density—1 lb/gal = 7.48 lb/ft^3 = 42 lb/bbl

Basic Conversion Factors

Length— 1 m = 3.281 ft = 39.37 in.
1 ft = 0.305 m = 30.5 cm = 3,050 mm
1 mile = 1.61 km
1 km = 0.621 mile

Area— $1 m^2 = 10.76 ft^2 = 1,549 in.^2$
$1 ft^2 = 0.0929 m^2 = 929.4 cm^2$

Volume—$1 m^3 = 35.32 ft^3 = 6.29 bbl$
$1 l = 0.035 ft^3 = 61 in.^3$
$1 ft^3 = 0.0283 m^3 = 28.3 l$
$1 bbl = 0.159 m^3 = 159 l$

Mass— 1 kg = 2.205 lb
1 lb = 0.454 kg = 454 g
1 metric ton = 1,000 kg = 2,205 lb

Density—$1 kg/m^3 = 0.0624 lb/ft^3$
$1 lb/ft^3 = 16.02 kg/m^3 = 0.01602 g/cm^3$
$1 g/cm^3 = 62.4 lb/ft^3$

Force— 1 kg force = 2.205 lb force
1 lb force = 0.454 kg force

Work &
Heat— 1 BTU = 0.252 kilocalories (kcal)
1 kcal = 3.97 BTU

Power— 1 kilowatt (kw) = 860 kcal/hr = 3,415 BTU/hr
= 1.341 horsepower (hp)
1 hp = 0.746 kw = 641 kcal/hr = 2,545 BTU/hr

Enthalpy—1 kcal/kg = 1.8 BTU/lb
1 BTU/lb = 0.556 kcal/kg

Pressure—1 bar = $14.51 lb/in.^2$ (psi) = 0.987 atmospheres
(atm) = $1.02 kg/cm^2$
$1 kg/cm^2 = 14.22 psi = 0.968 atm$
$1 psi = 0.0703 kg/cm^2$

Temperature—°C = 0.556 (°F − 32)
°K = °C + 273
°F = 1.8 °C + 32
°R = °F + 460

ADDENDUM

ADDENDUM